Lecture Notes in Mathematics

Edited by A. Dold, Heidelberg and B. Eckmann, Zürich

T0233424

356

W. L. J. van der Kallen

Mathematisch Instituut der Rijksuniversiteit Utrecht
De Uithof, Utrecht/Netherlands

Infinitesimally Central Extensions of Chevalley Groups

Springer-Verlag
Berlin · Heidelberg · New York 1973

AMS Subject Classifications (1970): Primary: 20G10, 20G15, 17B45, 17B55
Secondary: 20G05, 20H15, 50B30, 17B20, 20F25

ISBN 3-540-06559-8 Springer-Verlag Berlin · Heidelberg · New York
ISBN 0-387-06559-8 Springer-Verlag New York · Heidelberg · Berlin

Offsetdruck: Julius Beltz, Hemsbach/Bergstr.

Contents

Conventions		1
Section 1	Universal central extensions. Central trick	2
Section 2	Degenerate sums. The extension $r : \underline{g}'_{\mathbb{Z}} \to \underline{g}_{\mathbb{Z}}$	5
Section 3	The action $\hat{\text{Ad}}$. Structure of $\underline{g}^*_{\mathbb{Z}}$, $\underline{g}^*_R$	22
Section 4	Admissible lattices and the category $\mathcal{L}_V$ Σ-connected components	39
Section 5	The G-module ker π	46
Section 6	G-invariant [p]-structures	51
Section 7	The extension $\phi : G^* \to G$	54
Section 8	Extensions of G by a G-module	59
Section 9	The Hochschild groups	61
Section 10	The existence of $\phi : G^* \to G$	65
Section 11	Relations in the open cell	97
Section 12	Representatives in G* of the Weyl group	119
Section 13	The Theorem of generators and relations and its consequences	124
Section 14	The group functor G*	138
References		141
List of Notations		144
Index		147

Introduction

In these notes we study the connection between infinitesimally central extensions of Chevalley groups and universal central extensions of their Lie algebras. Here an infinitesimally central extension is a morphism of algebraic groups $\phi : H \to G$ such that, if $\underline{g}$, $\underline{h}$ denote the Lie algebra of G, H respectively,

(i) ϕ is surjective and separable,

(ii) the kernel of the derivative $d\phi$ of ϕ is contained in the centre of the Lie algebra $\underline{h}$.

We will restrict ourselves to the case that $\underline{h} = [\underline{h},\underline{h}]$.

Assume that $\underline{g} = [\underline{g},\underline{g}]$. Then a universal central extension $\pi : \underline{g}^* \to \underline{g}$ exists. It may be characterized as a homomorphism $\pi : \underline{g}^* \to \underline{g}$ such that

(i) π is surjective,

(ii) $\underline{g}^* = [\underline{g}^*,\underline{g}^*]$,

(iii) the kernel of π is contained in the centre of $\underline{g}^*$,

(iv) $\underline{g}^*$ is universal with respect to (i), (ii), (iii).

Condition (iv) is equivalent to

(iv)' If $\tau : \underline{g}' \to \underline{g}^*$ is a homomorphism satisfying (i), (ii), (iii) with π replaced by τ and $\underline{g}^*$ replaced by $\underline{g}'$, then τ is an isomorphism. (See section 1 of these notes or [22]).

Let G be a Chevalley group with Lie algebra $\underline{g}$ such that $\underline{g} = [\underline{g},\underline{g}]$. If the characteristic is not 2 or 3 then the universal central extension $\pi : \underline{g}^* \to \underline{g}$ is trivial, i.e. π is an isomorphism. This was proved by Steinberg in [23]. In section 3 we complete this result. We determine the structure of $\underline{g}^*$ in arbitrary characteristic by solving the analogous problem over $\mathbb{Z}$. (see Theorem 3.5 and Proposition 1.3 (vi)).

In describing $\underline{g}^*$ the notion of a degenerate sum in the lattice spanned by a root system is helpful. A degenerate sum is a sum of two linearly independent roots which is itself a p-multiple of a weight. (p is the characteristic). These degenerate sums are classified in section 2. It is seen that they only occur in characteristics 2 and 3. If there are no degenerate sums then $\underline{g}^* = \underline{g}$. (This generalizes Steinbergs result).

Let $\phi : H \to G$ be an infinitesimally central extension with $\underline{h} = [\underline{h},\underline{h}]$. Then $\underline{h}$ is isomorphic to a quotient of $\underline{g}^*$. If $\underline{g} = \underline{g}^*$ then $\underline{h} = \underline{g}$ and the connected component of H is a quotient of the simply connected covering of G. (See Springer-Steinberg, [2] E, §2). So the simply connected covering is a universal element in the class of extensions under consideration. Now we assume that $\underline{g} \neq \underline{g}^*$. Then we look for an extension $\phi : H \to G$ as above such that $\underline{h}$ is isomorphic to $\underline{g}^*$ and we ask whether this extension is a universal element. The existence of an extension with $\underline{h} \simeq \underline{g}^*$ is proved in section 10 for a simply connected almost simple Chevalley group G. The proof is based on the construction (case by case) of a suitable 2-cocycle of G in ker π. (There is a natural action of G on $\underline{g}^*$ which gives ker π the structure of a G-module). One gets a Hochschild-extension $\phi : G^* \to G$ which satisfies the requirements. Note that its radical is isomorphic to ker π and is hence commutative. Now we deal with the question whether ϕ is universal in the class of infinitesimally central extensions $H \to G$ with $\underline{h} = [\underline{h},\underline{h}]$. The answer is affirmative if G is not of type B_3 in characteristic 2. (In the case of type B_3 in characteristic 2 the class also contains extensions with non-commutative radicals. We don't give a proof of this fact). More generally, if G is Chevalley group with $\underline{g} = [\underline{g},\underline{g}]$ and if G has no factor of type B_3 in characteristic 2, then there is a universal solution $\phi_1 : G_1^* \to G$. (see Theorem 13.9). It is obtained by applying the solutions from section 10 to the simply connected coverings of the almost simple factors of G. Here it should be noted that $\underline{g} \neq \underline{g}^*$ implies that G has a simply connected factor (see 7.1, Remark). The proof of the fact that ϕ_1 is universal resembles the proof of the "Théorème fondamental" in [12], Exposé XXIII: We construct a set of generators and defining relations for G_1^* and prove that the same relations hold in all extensions of the class under consideration. (They are not <u>defining</u> relations for all these extensions). The generators and relations are very similar to Steinbergs generators and defining relations for a simply connected Chevalley group (see [22] or [23]). The analogy with simply connected Chevalley groups is also stressed by results about the group of automorphisms of G^* (see Corollary 13.7) and about embeddings of groups of distinct types into each other (see Theorem 13.14 and compare with [22] or [24]).

I feel indebted to professor T.A. Springer for his frequent advice and to professor F.D. Veldkamp who suggested the search for the groups G^*. I owe much to Mark Krusemeyer and Roelof Bruggeman for many useful discussions. I wish to thank miss A. van Hoof and mrs. P. van der Kuilen for careful typing.

CONVENTIONS

We will use mainly the same terminology as Borel in [1] and Steinberg in [22]. There are some modifications:

1. All algebraic groups are assumed to be affine.

2. All Chevalley groups are considered as algebraic groups. So a Chevalley group is an algebraic group that is obtained by the Chevalley construction from a faithful representation of a complex semi-simple Lie algebra. It is not necessarily of adjoint type. In fact we shall usually consider the simply connected types.

In dealing with varieties (not necessarily irreducible), we shall, as usual, write V for the set V(K) (or V_K) of K-rational points in V, K being an algebraically closed field. A map $V \to W$ shall be called a morphism, if it is a morphism of varieties.

3. In order to avoid ambiguities, a morphism of algebraic groups will be called a homomorphism and not just a morphism. So we shall speak of morphisms between algebraic groups that are not homomorphisms, but just morphisms of varieties.

4. If only one root length occurs in a root system then all roots are called long and not short.

§1. Universal central extensions. Central trick

In this section we introduce universal central extensions of Lie algebras, cf. [22]).

1.1. Let R be a ring. (Rings are commutative and have a unit).

A *Lie algebra* over R is an R-module $\underline{g}$, together with an R-bilinear composition

$[\ ,\ ] : \underline{g} \times \underline{g} \to \underline{g}$ that satisfies

(i) $[X,X] = 0$ for all $X \in \underline{g}$ (anti-symmetry).

(ii) $[X,[Y,Z]] + [Z,[X,Y]] + [Y,[Z,X]] = 0$ for all $X,Y,Z \in \underline{g}$. (Jacobi-relation).

So a Lie algebra over R is not necessarily a free R-module.

Homomorphisms are defined as usual. The *centre* of $\underline{g}$, i.e. $\{X \in \underline{g} | [X,Y] = 0 \text{ for all } Y \in \underline{g}\}$, is denoted $\underline{z}(\underline{g})$. An *extension* of $\underline{g}$ is a surjective homomorphism of Lie algebras $\pi: \underline{k} \to \underline{g}$. A *central extension* is an extension $\pi:\underline{k}\to\underline{g}$, satisfying $\ker \pi \subset \underline{z}(\underline{k})$.
A *universal central extension* is a central extension $\pi:\underline{g}^* \to \underline{g}$ with the property:
If $\phi : \underline{k} \to \underline{g}$ is a central extension, then there is exactly one homomorphism $\psi: \underline{g}^* \to \underline{k}$ such that $\phi\circ\psi = \pi$. Note that ψ is not necessarily surjective. Henceforth $\pi: \underline{g}^* \to \underline{g}$ will always denote a universal central extension of $\underline{g}$. A Lie algebra $\underline{g}$ is *centrally closed* if $\text{id}: \underline{g} \to \underline{g}$ is a universal central extension.

1.2. LEMMA (*central trick*).
If $\pi: \underline{k} \to \underline{g}$ *is a central extension, and if* $X,X',Y,Y' \in \underline{k}$ *are such that* $\pi X = \pi X'$ *and* $\pi Y = \pi Y'$, *then* $[X,Y] = [X',Y']$.

PROOF. $Y-Y' \in \ker \pi \subset \underline{z}(\underline{k})$, so $[X,Y] = [X,Y']$. In the same way $[X,Y'] = [X',Y']$, whence the lemma.

The central trick is an important tool for lifting properties from $\underline{g}$ to $\underline{k}$. Its usefulness was demonstrated by R. Steinberg in [23].

1.3. PROPOSITION. (cf. [22], §7).

(i) If $\phi: \underline{g}' \to \underline{g}$ and $\psi: \underline{g}'' \to \underline{g}'$ are central extensions, and $[\underline{g}'', \underline{g}''] = \underline{g}''$, then $\phi \circ \psi: \underline{g}'' \to \underline{g}$ is a central extension.

(ii) $\underline{g}$ has a universal central extension if and only if $\underline{g} = [\underline{g}, \underline{g}]$.

(iii) Universal central extensions of $\underline{g}$ are isomorphic.

(iv) If $\pi: \underline{g}^* \to \underline{g}$ is a universal central extension, then $[\underline{g}^*, \underline{g}^*] = \underline{g}^*$ and $\underline{g}^*$ is centrally closed.

(v) If $\pi: \underline{g}^* \to \underline{g}$ is a universal central extension, $\psi: \underline{g} \to \underline{k}$ a homomorphism, $\phi: \underline{k}' \to \underline{k}$ a central extension, then there is exactly one $\hat{\psi}: \underline{g}^* \to \underline{k}'$ such that $\phi \circ \hat{\psi} = \psi \circ \pi$.
If ψ is surjective then $\hat{\psi}(\underline{g}^*) = [\underline{k}', \underline{k}']$.

(vi) Let R,S be rings, S an R-algebra.
Let $\underline{g}$ be a Lie algebra over R with universal central extension $\pi: \underline{g}^* \to \underline{g}$.
Then $\pi \otimes \mathrm{id}: \underline{g}^* \otimes_R S \to \underline{g} \otimes_R S$ is a universal central extension of Lie algebras over S.

PROOF.

(i) From Jacobi it follows that
$[\ker(\phi\psi), [\underline{g}'', \underline{g}'']] \subset [\underline{g}'', [\ker(\phi\psi), \underline{g}'']]$.
And
$\psi[\ker(\phi\psi), \underline{g}''] \subset [\ker \phi, \underline{g}'] = 0$, so
$[\underline{g}'', [\ker(\phi\psi), \underline{g}'']] \subset [\underline{g}'', \ker \psi] = 0$.

(ii) Only if part.
Set r = projection of $\underline{g}$ on $\underline{g}/[\underline{g}, \underline{g}]$. Suppose $\pi: \underline{g}^* \to \underline{g}$ exists.

If $\sigma: A \to B$ and $\tau: A \to C$, then we denote $\sigma \oplus \tau$ the map $x \mapsto (\sigma(x), \tau(x))$. So we have $\pi \oplus r\pi: \underline{g}^* \to \underline{g} \oplus \underline{g}/[\underline{g},\underline{g}]$ and $\pi \oplus 0: \underline{g}^* \to \underline{g} \oplus \underline{g}/[\underline{g},\underline{g}]$. The projection of $\underline{g} \oplus \underline{g}/[\underline{g},\underline{g}]$ on the first factor is a central extension p_1 of $\underline{g}$. As $p_1(\pi \oplus r\pi) = p_1(\pi \oplus 0)$, we have $r\pi = 0$ by unicity, so $g = [\underline{g},\underline{g}]$.

If part.

We give a construction of $\pi: \underline{g}^* \to \underline{g}$, supposing that $\underline{g} = [\underline{g},\underline{g}]$.

In the R-module $\underline{g} \otimes_R \underline{g}$ we define the bilinear composition $[\ ,\ ]$ by $[X \otimes Y, X' \otimes Y'] = [X,Y] \otimes [X',Y']$.

Let N be the submodule generated by

(1) $[P,P]$,

(2) $[P,[Q,R]] + [R,[P,Q]] + [Q,[R,P]]$, $(P,Q,R \in \underline{g} \otimes \underline{g})$, and put $\underline{g}^* = \underline{g} \otimes_R \underline{g}/N$. Then $\underline{g}^*$ is a Lie algebra.

Choose $\pi: \underline{g}^* \to \underline{g}$ such that $\pi\{X \otimes Y\} = [X,Y]$. (Here $\{X \otimes Y\}$ denotes the residue class of $X \otimes Y$).

It is easy to check that π is well-defined. Then it is seen from $\underline{g} = [\underline{g},\underline{g}]$ that π is an extension, which is central because of the definition of $[\ ,\ ]$ in $\underline{g} \otimes \underline{g}$.

Now let $\phi: \underline{k} \to \underline{g}$ be a central extension.

Choose a section s of ϕ, i.e. a mapping s such that $\phi \circ s = \mathrm{id}$. Using the central trick (Lemma 1.2), we see that $(X,Y) \mapsto [sX,sY]$ is bilinear, so a mapping $\underline{g} \otimes \underline{g} \to \underline{k}$ is induced. Using the central trick again, we see that it is a homomorphism of non-associative algebras. Therefore a Lie algebra homomorphism $\psi: \underline{g}^* \to \underline{k}$ is induced, satisfying $\phi \circ \psi\{X \otimes Y\} = \phi[sX,sY] = [X,Y] = \pi\{X \otimes Y\}$.

Now suppose ψ' is a homomorphism satisfying $\phi \circ \psi' = \pi$. Then $\psi'[P,Q] = [\psi'P, \psi'Q] = [\psi P,\psi Q] = \psi[P,Q]$ by the central trick. As

$\underline{g} = [\underline{g}, \underline{g}]$, we see that every $\{X \otimes Y\} \in \underline{g}^*$ is of the form $[P,Q]$.

So we are done, and we also have proved that $\underline{g}^* = [\underline{g}^*, \underline{g}^*]$.

(iii) Use abstract nonsense.

(iv) By the last remark in the proof of (ii) we only have to prove that $\underline{g}^*$ is centrally closed. Let $\underline{g}^{**} \to \underline{g}^*$ be a universal central extension. Using (i) we see that $\underline{g}^{**} \to \underline{g}$ is a central extension. So the extension $\underline{g}^{**} \to \underline{g}^*$ splits, and we have $\underline{g}^{**} \cong \underline{g}^* \oplus \underline{i}$ where $\underline{i}$ denotes the abelian Lie algebra ker $(\underline{g}^{**} \to \underline{g}^*)$. As $\underline{g}^{**} = [\underline{g}^{**}, \underline{g}^{**}]$ this implies that $\underline{i} = 0$.

(v) As in the proof of (iv) we choose a section s of ϕ and see that $(X,Y) \mapsto [s\psi X, s\psi Y]$ is bilinear. Again a Lie algebra homomorphism is induced, and again it is unique by the central trick. Now suppose ψ is surjective. Then $\hat{\psi}\underline{g}^* = \hat{\psi}[\underline{g}^*, \underline{g}^*] = [\hat{\psi}\underline{g}^*, \hat{\psi}\underline{g}^*] = [\underline{k}', \underline{k}']$ by the central trick.

(vi) The construction of $\underline{g}^*$ we gave in the proof of (ii) commutes with the base transformation from R to S. (The functor $- \otimes_R S$ is right exact).

§2. Degenerate sums. The extension $r : \underline{g}'_{\mathbb{Z}} \to \underline{g}_{\mathbb{Z}}$.

In this section we shall introduce degenerate sums. Besides that we derive some technicalities involving root systems and their classification (see [4]). Omitting some defining relations for $\underline{g}_{\mathbb{Z}}$ we construct a central extension $r : \underline{g}'_{\mathbb{Z}} \to \underline{g}_{\mathbb{Z}}$.

2.1. NOTATIONS.

We are going to consider Lie algebras of simply connected almost simple Chevalley groups in characteristic $p > 0$. So let k be a field of characteristic $p > 0$, K its algebraic closure, G a simply connected almost simple Chevalley group viewed as an algebraic group defined over k, $\underline{g}$ the Lie algebra of G. The set of k-rational points in $\underline{g}$ is denoted $\underline{g}_k$.

It is a Lie algebra over k. We use the following standard notations (see [22]).

l = rank of G,

$\underline{g}_{\mathbb{C}}$ = the complex Lie algebra corresponding to $\underline{g}$,

Σ = the (irreducible) root system. (It is assumed to be ordered).

W = Weyl group,

$\{X_\alpha, H_\alpha | \alpha \in \Sigma\}$ = set of Chevalley generators in $\underline{g}_{\mathbb{C}}$, or the corresponding set of generators in $\underline{g}$.

$\{N_{\alpha\beta}\}$ = the corresponding set of structure constants,

$\{x_\alpha(t) | \alpha \in \Sigma, t \in K\}$ = the set of generators of G.

$w_\alpha(t) = x_\alpha(t)\, x_{-\alpha}(-t^{-1})\, x_\alpha(t)$, for $\alpha \in \Sigma$, $t \in K^\times = K\setminus\{0\}$.

$h_\alpha(t) = w_\alpha(t)\, w_\alpha(1)^{-1}$, for $\alpha \in \Sigma$, $t \in K^\times$.

Ω = the open cell, consisting of the elements

$$\prod_{\alpha<0} x_\alpha(u_\alpha) \prod_{\alpha \text{ simple}} h_\alpha(t_\alpha) \prod_{\alpha>0} x_\alpha(u_\alpha), \text{ where}$$

$u_\alpha \in K$, $t_\alpha \in K^\times$ (see [8], Proposition 1).

$(x,y) = xyx^{-1}y^{-1}$ if x,y are group elements,

(x,y) = the inner product of x and y if x,y are elements of a real vector space.

The notation may also be used for an element of a direct product of varieties.

$\underline{g}_{\mathbb{Z}}$ = $\mathbb{Z}$ - Lie algebra generated by the X_α, H_α in $\underline{g}_{\mathbb{C}}$.

Γ = lattice of weights,

Γ_0 = sublattice generated by the roots.

$\langle\alpha,\beta\rangle = \dfrac{2(\alpha,\beta)}{(\beta,\beta)}$ for α,β in the real vector space with inner product which is generated by Σ. ($\beta \neq 0$).

If $\alpha,\beta \in \Sigma$, then $\langle\alpha,\beta\rangle \in \mathbb{Z}$. If α,β are (linearly) independant roots with $|\alpha| \leqslant |\beta|$ then $|<\alpha,\beta>| \leqslant 1$.

$\{\alpha_1,\ldots,\alpha_l\}$ = set of simple roots, numbered as in [4],

$\{\varepsilon_i\}$ = orthonormal basis that is used in [4] to describe the root system,

$\{\delta_i\}$ = set of fundamental weights.

So we have $\underline{g} \cong \underline{g}_{\mathbb{Z}} \otimes K$,

$\Gamma = \{\alpha | <\alpha,\Sigma> \subset \mathbb{Z}\}$.

The ordering of Σ induces an ordering of Γ defined by: $\alpha \geqslant \beta$ if $\alpha-\beta$ is a positive linear combination of the simple roots (see [4], Ch. VI §1.6).

2.2. PROPOSITION. $\underline{g}_k = [\underline{g}_k, \underline{g}_k]$ *if and only if* $\Sigma \cap p\Gamma = \emptyset$.

PROOF. $\Sigma \cap p\Gamma$ consists of those roots α for which $<\alpha,\Sigma> \subset p\,\mathbb{Z}$. So if $\Sigma \cap p\Gamma = \emptyset$, then for every $\alpha \in \Sigma$ there is $\beta \in \Sigma$ such that $X_\alpha = <\alpha,\beta>^{-1}[H_\beta, X_\alpha]$ in $\underline{g}_k$. The elements X_α generate $\underline{g}_k$ as a Lie algebra. Conversely, suppose $\Sigma \cap p\Gamma \neq \emptyset$. Since $\Sigma \cap p\Gamma$ consists of W-orbits, it contains a simple root α. For all simple roots β one has $<\alpha,\beta> \in p\,\mathbb{Z}$. Taking $\beta = \alpha$ one sees $p = 2$. Taking roots corresponding to neighbours in the Dynkin diagram for β, one concludes that Σ is of type C_l, $l \geqslant 1$ ($C_1 = A_1$, $C_2 = B_2$). One now checks that $\underline{g}_k \neq [\underline{g}_k, \underline{g}_k]$ in these cases (see [17], Lemma 7).

COROLLARY.

(i) $\underline{g}_k = [\underline{g}_k, \underline{g}_k]$ *if and only if* Σ *is not of type* C_l ($l \geqslant 1$), *or* $p > 2$.

(ii) $\underline{g}_{\mathbb{Z}} = [\underline{g}_{\mathbb{Z}}, \underline{g}_{\mathbb{Z}}]$ *if and only if* Σ *is not of type* C_l ($l \geqslant 1$).

PROOF. We have to prove (ii).

If part.

For every p we have $(g_{\mathbb{Z}} \bmod [g_{\mathbb{Z}}, g_{\mathbb{Z}}]) \otimes_{\mathbb{Z}} \mathbb{F}_p = 0$ by (i).
So $g_{\mathbb{Z}} \bmod [g_{\mathbb{Z}}, g_{\mathbb{Z}}] = 0$

Only if part.

Take p = 2 and use (i).

2.3. LEMMA. Let α,β be independent roots.
Then there is $\gamma \in \Sigma$ such that $\Sigma_1 = (\mathbb{Q}\alpha+\mathbb{Q}\beta+\mathbb{Q}\gamma) \cap \Sigma$ is an irreducible root system.
If rank $\Sigma > 2$ then γ may be chosen such that rank $\Sigma_1 = 3$.

PROOF.

Let $\lambda_0,\dots,\lambda_q$ be a sequence of roots such that $\lambda_0 = \alpha$, $(\lambda_i,\lambda_{i+1})\neq 0$, $\lambda_q = \beta$. Such a sequence exists because Σ is irreducible. Now suppose q is minimal and $q > 2$. As $<\lambda_1, \lambda_2> \neq 0$, we have $\lambda_1 - \lambda_2 \in \Sigma$ or $\lambda_1 + \lambda_2 \in \Sigma$. Say $\lambda_1 + \lambda_2 \in \Sigma$. As q is minimal, we have $(\lambda_2,\lambda_0) = 0$. And $(\lambda_0, \lambda_1) \neq 0$, so $(\lambda_1 + \lambda_2, \lambda_0) \neq 0$. In the same way $(\lambda_1 + \lambda_2, \lambda_3) \neq 0$. But then $\lambda_0, \lambda_1 + \lambda_2, \lambda_3, \dots, \lambda_q$ is a shorter sequence, which is a contradiction. So we may take $q \leqslant 2$. Define $\gamma = \lambda_1$. Then every irreducible component of Σ_1 which contains γ contains α and β. So Σ_1 is irreducible. If rank $\Sigma > 2$, then we have to consider two cases:

First suppose $\Sigma_2 = (\mathbb{Q}\alpha + \mathbb{Q}\beta) \cap \Sigma$ is a reducible root system (i.e. of type $A_1 \times A_1$). Then we choose γ as above.

Secondly suppose Σ_2 is irreducible.

Then we choose $\gamma \in \Sigma$ such that γ is not orthogonal to Σ_2 and γ is not in Σ_2. We always get an irreducible Σ_1 of rank 3 this way.

2.4. DEFINITION. Let $\gamma \in \Gamma$, $n \in \mathbb{Z}$, $n > 1$. Then γ is called a *degenerate sum with respect to n* if

(i) There are independent roots α,β with $\alpha + \beta = \gamma$.

(ii) $\gamma \in n\Gamma$. This means that $\langle\gamma,\Sigma\rangle \subset n\mathbb{Z}$.

If $n = p$ then we just say that γ is a degenerate sum, or that γ is degenerate.

2.5. LEMMA. *Let* Σ *be as above*, Σ_1 *a subset of* Σ. *If* Σ_1 *is an irreducible root system, and* $\alpha,\beta \in \Sigma_1$ *are independent, such that* $\alpha + \beta$ *is degenerate with respect to* n *in* Σ, *then* $\alpha + \beta$ *is a degenerate sum with respect to* n *in* Σ_1 *too*.

The proof is trivial.

REMARK. The converse does not hold, as one can see from 2.8., Table 1.

2.6. LEMMA.

(i) *If* $n > 3$ *then no degenerate sums with respect to* n *exist. So degenerate sums may only occur if* $p = 2$ *or* $p = 3$.

(ii) *If* $p = 2$, $\alpha,\beta \in \Sigma$ *are independent*, $\alpha + \beta$ *is degenerate, then* $(\alpha,\beta) = 0$.

(iii) *If* $\alpha,\beta,\gamma,\delta$ *are distinct roots, while* $\alpha + \beta = \gamma + \delta$ *is a degenerate sum, then* $p = 2$ *and the same root lengths occur in both pairs of roots*.

PROOF. If α,β are independent, $|\alpha| \leqslant |\beta|$, then $0 < 2 + \langle\alpha,\beta\rangle < 4$, so that $1 \leqslant \langle\alpha+\beta,\beta\rangle \leqslant 3$. This implies (i). If furthermore $\langle\alpha+\beta,\beta\rangle \in 2\mathbb{Z}$ then it follows that $\langle\alpha,\beta\rangle = 0$, whence (ii).

(iii) Let β have a largest length in the set $\{\alpha,\beta,\gamma,\delta\}$. Then $|\langle\gamma+\delta,\beta\rangle| \leqslant 2$.

And $\langle\gamma+\delta,\beta\rangle = \langle\alpha+\beta,\beta\rangle$ is again strictly positive.

So $p = 2$ and $\alpha \perp \beta$, $\gamma \perp \delta$.

So $|\alpha|^2 + |\beta|^2 = |\alpha+\beta|^2 = |\gamma+\delta|^2 = |\gamma|^2 + |\delta|^2$.

As, for fixed Σ, there are at most two possibilities for the values of the root lengths, there are at most four possibilities for the value of

$|\alpha'|^2 + |\alpha''|^2$, α', $\alpha'' \in \Sigma$.

These values correspond to the occurrence of root lengths in the pair α', α''.

2.7. We are now going to classify degenerate sums. We may restrict ourselves to one representative for each orbit under the action of W. Results will be given in 2.8., Table 1.

EXPLICIT DETERMINATION.

According to lemma 2.6. we may restrict ourselves to $p = 2$ and $p = 3$. First let $p = 3$.

Choose a normalisation of the inner product such that the shortest roots have lengths 1. Recall that Γ_0 is the lattice generated by Σ. For $\gamma \in \Gamma_0$ we have $(\gamma,\gamma) \in \mathbb{Z}$. Set n = order of Γ/Γ_0 (= "indice de connexion"). (See [4]). Then $n^2\ (\gamma,\gamma) \in \mathbb{Z}$ for every $\gamma \in \Gamma$.

Now let $\alpha,\beta \in \Sigma$ with $\alpha+\beta$ degenerate, $|\alpha| \leqslant |\beta|$. Then $\alpha+\beta \in 3\Gamma$, so $n^2(\alpha+\beta,\ \alpha+\beta) \in 9\mathbb{Z}$. And $(\alpha+\beta,\ \alpha+\beta) = (\alpha,\alpha) + \langle\alpha,\beta\rangle(\beta,\beta)+(\beta,\beta) \leqslant$ $\leqslant 3 + 3 + 3 = 9$. So either n is divisible by 3 or $(\alpha+\beta,\ \alpha+\beta) = 9$.

In the latter case, Σ is of type G_2 and α,β are two long roots making an angle $\pi/3$. This yields a degenerate sum indeed, because the sum is p times a root.

In the case that n is divisible by 3, Σ is of type A_{3m-1} or E_6.

So now we may assume that all root lengths are equal. As $\langle\alpha+\beta,\ \beta\rangle \in 3\mathbb{Z}$, we see that $\langle\alpha,\beta\rangle = 1$, which means that they make an angle $\pi/3$. In A_2 this yields two orbits of degenerate sums. Now suppose Σ is of type A_{3m-1}, $m > 1$, or E_6. Using lemma 2.3

we get a root system Σ_1, containing α and β, with rank $\Sigma_1 = 3$.
In this system $\alpha+\beta$ should be degenerate too. (lemma 2.5.).
But we have seen that no root system of rank 3 yields degenerate sums. So we are done for $p = 3$.

Now let $p = 2$. As we know from lemma 2.6, we have $\alpha \perp \beta$.
Consider $\Sigma_2 = (\mathbb{Q}\alpha+\mathbb{Q}\beta) \cap \Sigma$. It is a root system, so we can choose a system of simple roots in it, containing α (see [4], Ch. VI, §1, Prop. 15). (If possible, we choose this system of simple roots in such a way that β is simple too). According to [4], Ch. VII, §1, Prop. 24, there is a system of simple roots in Σ, containing the one chosen in Σ_2.
Now there are two possibilities:
1). Σ_2 is reducible.
In this case β has also been chosen to be simple, and we have to deal with Dynkin diagrams. Say $\alpha = \alpha_r$, $\beta = \alpha_s$, where $\alpha_1,\ldots,\alpha_l$ are the simple roots. As $(\alpha,\beta) = 0$, the points r and s are not neighbours in the Dynkin diagram.
So Σ has rank > 2. Now consider such a pair r, s in a Dynkin diagram, consisting of two points that are not neighbours. The fact that $\alpha+\beta$ is degenerate may be expressed by the relations
$\langle\alpha_r,\alpha_i\rangle \equiv \langle\alpha_s, \alpha_i\rangle \bmod 2$, $i = 1,\ldots,l$.
For $i = r$ and $i = s$ the relation is always satisfied and it is also satisfied if i is adjacent to neither r or s in the Dynkin diagram. So we have to look at neighbours of r and s. For a common neighbour i the relation is satisfied if and only if α_i has maximal lenght in the set

$\{\alpha_i, a_r, \alpha_s\}$. For other neighbours, say neighbours i of r that are not adjacent to s, the relation is equivalent to $(\alpha_r, \alpha_r) = 2(\alpha_i, \alpha_i)$. There is at most one place in a Dynkin diagram where $(\alpha_j, \alpha_j) = 2(\alpha_i, \alpha_i)$ is satisfied for neighbours i,j, so there is at most one non-common neighbour.

It is easily seen that these requirements for the behaviour of neighbours select one pair r,s if Σ is of type $A_3 = D_3$, $D_l (l \geqslant 3)$, B_3, B_4, and don't permit any pairs in other cases.

2). Σ_2 is irreducible.

As $(\alpha,\beta) = 0$ we have Σ_2 of type B_2 or G_2.

First let Σ_2 be of type G_2. Up to the action of the Weyl group, there is just one pair of orthogonal roots.

This pair α,β yields a sum that is twice a root. Hence it is a degenerate sum.

Now let Σ_2 be of type B_2.

There are two possibilities for an orthogonal pair:

Both roots are short or both roots are long. If they are short, their sum is a long root. So we have to do with the case $\Sigma \cap p\Gamma \neq \emptyset$. That is, Σ is of type C_l (see 2.2.).

A long root is degenerate in this case indeed. Finally, if both roots are long, their sum is twice a root, so it is a degenerate sum again. This situation occurs in B_l, C_l, F_4.

2.8. Summing up, we can list results as in Table 1. In this table all W-orbits of degenerate sums and of elements in $\Sigma \cap p\Gamma$ are given. A notation like $\alpha_1 + \alpha_3$ $[6, 2\delta_2]$ means that there is an orbit consisting of 6 elements, with $\alpha_1 + \alpha_3$ and $2\delta_2$ as representatives. The number 6 and the fundamental weight δ_2 are found with the help of the "Planches" in [4].

Table 1.

Type	Dynkin diagram	$\Sigma \cap p\Gamma$		Degenerate sums		
		p=2	p>2	p=2	p=3	p>3
A_1		$\alpha_1[2,2\delta_1]$	-	-	-	-
A_2	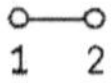	-	-	-	$\begin{cases}2\alpha_1+\alpha_2[3,3\delta_1]\\ 2\alpha_2+\alpha_1[3,3\delta_2]\end{cases}$	-
A_3	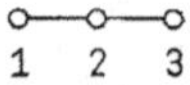	-	-	$\alpha_1+\alpha_3[6,2\delta_2]$	-	-
B_3	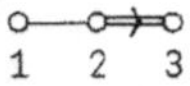	-	-	$\begin{cases}2\alpha_3[6,2\delta_1]\\ \alpha_1+\alpha_3[8,2\delta_3]\end{cases}$	-	-
B_4	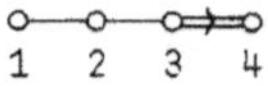	-	-	$\begin{cases}2\alpha_4[8,2\delta_1]\\ \alpha_1+\alpha_3[16,2\delta_4]\end{cases}$	-	-
$B_l\,(l>4)$	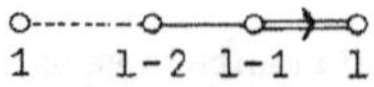	-	-	$2\alpha_1[2l,2\delta_1]$	-	-
$C_l\,(l\geqslant 2)$	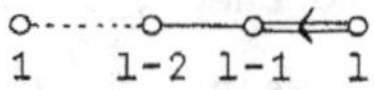	$\alpha_1[2l,2\delta_1]$	-	$\begin{cases}2\alpha_1[2l^2-2l,2\delta_2]\\ \alpha_1[2l,2\delta_1]\end{cases}$	-	-
D_4	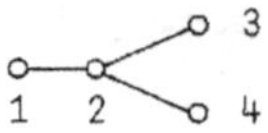	-	-	$\begin{cases}\alpha_1+\alpha_3[8,2\delta_4]\\ \alpha_1+\alpha_4[8,2\delta_3]\\ \alpha_3+\alpha_4[8,2\delta_1]\end{cases}$	-	-
$D_l\,(l>4)$	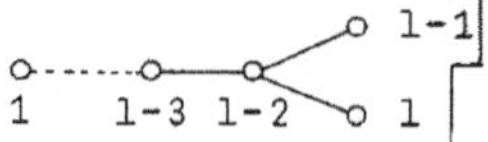	-	-	$\alpha_{l-1}+\alpha_l[2l,2\delta_1]$	-	-
F_4	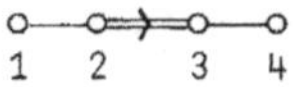	-	-	$2\alpha_3[24,2\delta_4]$	-	-
G_2		-	-	$2\alpha_1[6,2\delta_1]$	$3\alpha_1[6,3\delta_1]$	-
others	some	-	-	-	-	-

2.9. LEMMA.

(i) *If* γ *is a degenerate sum with respect to* p *then* $p^{-1}\gamma$ *is in the orbit of a fundamental weight.*

(ii) *If* Σ *is not of type* C_l, $l \geq 2$, *then the fundamental weight in (i) is a minimal dominant weight in the sense of the order defined in 2.1.*

(iii) *Let* Σ *be of a type such that degenerate sums with respect to* p *occur and let* α *be a short root. Then* $p\alpha$ *is a degenerate sum.*

PROOF.

(i) See Table 1.

(ii) Some cases are discussed in ([7], p. 20-03). Let δ be the fundamental weight that is found in (i). If it is not minimal, then there is a dominant weight α such that $\delta - \alpha > 0$. We may suppose that α is fundamental because fundamental weights are positive and α is a sum of them. It is easy to check, using the "Planches" of [4], that for each fundamental weight $\delta_i \neq \delta$ the difference $\delta - \delta_i$ is not positive. (Use the description of δ_i in terms of the α_j). Hence α does not exist, except in the cases C_l, where the check doesn't work.

(iii) See Table 1.

REMARK.

If the minimal dominant weight in (ii) is not a root, then it is a "Poids minuscule" in the sense of ([4], exercice 24, p. 226).

2.10. LEMMA.

If Σ has degenerate sums, then the order of Γ/Γ_0 is a power of p. In fact it is 1, p or p^2.

PROOF.

Compare Table 1 with the Planches again.

2.11. LEMMA.

Except for the cases B_3 and C_l, $l \geq 2$, all degenerate sums (in Γ) with respect to the same p have the same length.

2.12. PROPOSITION.

Let p be prime, $\gamma \in p\Gamma \cap \Gamma_0$, $\gamma \neq 0$.

Then γ is a degenerate sum with respect to p if and only if there is a long root α with $(\gamma,\gamma) \leq p(\alpha,\alpha)$.

PROOF.

Suppose γ is a degenerate sum. Choose a long root α such that $(\alpha,\gamma) > 0$. If $p = 3$ then it is seen from the table that $(\gamma,\gamma) = p(\alpha,\alpha)$.

If $p = 2$ then it follows from lemma 2.6, (ii) that $(\gamma,\gamma) \leq 2(\alpha,\alpha)$.

Conversely, suppose $(\gamma,\gamma) \leq p(\alpha,\alpha)$, $\gamma \in p\Gamma \cap \Gamma_0$, $\gamma \neq 0$. Recall that α is a long root such that $\langle\gamma,\alpha\rangle > 0$. Then $\langle\gamma,\alpha\rangle^2 \geq p^2$, so $\frac{4(\gamma,\gamma)(\alpha,\alpha)}{(\alpha,\alpha)^2} \geq p^2$, and hence $p(\alpha,\alpha) \geq (\gamma,\gamma) \geq \frac{p^2}{4}(\alpha,\alpha)$. It follows that $p \leq 4$, whence $p = 2$ or $p = 3$.

1). First suppose Σ is of type C_l. Then there is an orthogonal base (β_i), consisting of long roots. As $\langle\gamma,\beta_i\rangle \in p\mathbb{Z}$, we have $\gamma = \frac{p}{2} \sum_i n_i\beta_i$, $n_i \in \mathbb{Z}$.

So $(\gamma,\gamma) = \frac{p^2}{4} (\sum_i n_i^2) (\alpha,\alpha)$.

It follows that $\sum_i n_i^2 \leqslant \frac{4}{p}$.

There is no solution for $p = 3$, because $\gamma = \frac{3}{2} \beta_i$ is not in Γ_0.

For $p = 2$ there are two solutions, up to the action of the Weyl group. As there are also two orbits of degenerate sums, these solutions are degenerate sums.

2). From now on we exclude type $C_l (l \geqslant 1)$. First let $p=3$. Recall that α has been chosen such that $\langle\gamma,\alpha\rangle$ is strictly positive. The root α is the sum of two long roots. (Type C_l is excluded). Let β be one of them, such that $(\gamma,\beta) > 0$. Then we have: $\langle\gamma,\alpha\rangle \geqslant p$, $\langle\gamma,\beta\rangle \geqslant p$, and hence

$$0 \leqslant \frac{(\gamma-\alpha-\beta,\ \gamma-\alpha-\beta)}{(\alpha,\alpha)} = \frac{(\gamma,\gamma)}{(\alpha,\alpha)} + 2 - \langle\gamma,\alpha\rangle - \langle\gamma,\beta\rangle + 1 \leqslant$$

$p+2-p-p+1 = 0$.

It follows that $\gamma = \alpha + \beta$, hence γ is a degenerate sum.

Finally let $p = 2$.

We may suppose that γ is a dominant weight. Then $\gamma=2 \sum_i n_i\delta_i=\sum_i m_i\alpha_i$, where $m_i, n_i \in \mathbb{Z}$, $m_i \geqslant 0$, $n_i \geqslant 0$. (Recall that $\gamma \in p\Gamma \cap \Gamma_0$). As $\langle\gamma,\alpha_i\rangle \geqslant 0$ for each i, all m_i are strictly positive. (Consider an index in the Dynkin diagram adjacent to an index i where $m_i>0$). Now $2n_i = \langle\gamma,\alpha_i\rangle = 2m_i + \sum_{j\neq i} m_j \langle\alpha_j,\alpha_i\rangle < 2m_i$. So

(1) $m_i \geqslant n_i+1$.

Hence

(2) $2(\alpha,\alpha) \geqslant (\gamma,\gamma) = \sum_i m_i n_i(\alpha_i,\alpha_i) \geqslant \sum_i (n_i+1)\, n_i(\alpha_i,\alpha_i)$.

Suppose there are two indices r and s such that $n_r > 0$, $n_s > 0$. Then $2(\alpha,\alpha) \geqslant 2n_r(\alpha_r,\alpha_r) + 2n_s(\alpha_s,\alpha_s)$. It follows that α_r and α_s are short, so Σ is of type F_4. (Again we use that type C_l is excluded). Say s is the one that has two neighbours in the Dynkin diagram. As $m_r \geq 2$, we have

$2n_s = \langle\gamma,\alpha_s\rangle = 2m_s + \sum_{j\neq s} m_j \langle\alpha_j,\alpha_s\rangle < 2m_s - 2$.

Hence $m_s > 2$, and it follows from (2) that $2(\alpha,\alpha) \geq 3(\alpha_s,\alpha_s)+2(\alpha_r,\alpha_r)$, which is nonsense.

We may conclude that there is only one index r such that $n_r > 0$.

Suppose $n_r > 1$. Then $2(\alpha,\alpha) \geq (\gamma,\gamma) = m_r n_r(\alpha_r,\alpha_r) \geq 6(\alpha_r,\alpha_r)$ (see (1) and (2)).

So Σ is of type G_2, $m_r n_r = 6$, $m_r = 3$, $n_r = 2$. This is nonsense, because δ_r is a root in case G_2. What is left is the case $\gamma = 2\delta_r$. Then we have $\gamma/2 = \sum_i \frac{m_i}{2} \alpha_i$, $m_i \in \mathbb{Z}$, $m_r(\alpha_r,\alpha_r) \leq 2(\alpha,\alpha)$. All we have to do now, is to look in the Bourbaki Planches for such fundamental weights δ_r. For each type, there are as many of them as there are orbits of degenerate sums.

REMARK. In fact the proof gives another method to classify degenerate sums in characteristic 2. It also explains Lemma 2.9.(i), in characteristic 2.

2.13. The Lie algebra $\underline{g}_\emptyset$ is defined as a vector space by the following generators and relations:

Generators: X_α, $H_\alpha(\alpha \in \Sigma)$.

Relations:

(1) $H_\alpha + H_{-\alpha} = 0$ for $\alpha \in \Sigma$.

(2) $H_\alpha + \frac{(\beta,\beta)}{(\alpha,\alpha)} H_\beta + \frac{(\gamma,\gamma)}{(\alpha,\alpha)} H_\gamma = 0$ for $\alpha,\beta,\gamma \in \Sigma$, $\alpha+\beta+\gamma=0$,

$(\alpha,\alpha) \leq (\beta,\beta)$, $(\alpha,\alpha) \leq (\gamma,\gamma)$.

These relations follow from the fact that the left hand sides act trivially on roots. Every H_α can be expressed by means of relations (1), (2) in terms of the H_{α_i}. (α_i simple). So relations (1), (2) are sufficient to define $\underline{g}_\emptyset$, for reasons of dimension.

For α,β,γ as in (2), we have the Jacobi identity

$[X_\alpha, [X_\beta, X_\gamma]] + [X_\beta, [X_\gamma, X_\alpha]] + [X_\gamma, [X_\alpha, X_\beta]] = 0$,

which yields:

(3) $N_{\beta\gamma}H_\alpha + N_{\gamma\alpha}H_\beta + N_{\alpha\beta}H_\gamma = 0$.

As $\beta+\gamma = -\alpha$ is a root, $N_{\beta,\gamma} \neq 0$ and H_β, H_γ are linearly independent. So relation (3) is obtained from relation (2) by multiplying with the nonzero factor $N_{\beta,\gamma}$.

2.14. DEFINITION.

Let $\underline{g}'_{\mathbb{Z}}$ be the $\mathbb{Z}$ - module with generators X_α, $H_\alpha (\alpha \in \Sigma)$ and relations (1), (3) of 2.13. (So relation (2) is omitted).

We define the bilinear anti-symmetric composition [,] on $\underline{g}'_{\mathbb{Z}}$ by the usual relations:

$[X_\alpha, X_\beta] = N_{\alpha\beta}\ X_{\alpha+\beta}$ if $\alpha+\beta \in \Sigma$.

$[X_\alpha, X_{-\alpha}] = H_\alpha$.

$[X_\alpha, X_\beta] = 0$ if $\alpha+\beta \notin \Sigma \cup (0)$.

$[H_\alpha, X_\beta] = \langle\beta,\alpha\rangle X_\beta$.

$[H_\alpha, H_\beta] = 0$.

It is easily seen that this composition is well-defined. We now claim that $\underline{g}'_{\mathbb{Z}}$ is a Lie algebra. We only have to check Jacobi relations for the generators.

If $\alpha,\beta,\gamma \in \Sigma$, $\alpha+\beta+\gamma = 0$, then the Jacobi relation for X_α, X_β, X_γ is just relation (3) of 2.13. For other combinations of the generators the three terms in the Jacobi relation are multiples of one generator. So for those combinations the Jacobi relation follows from the fact that we use the same structure constants in $\underline{g}'_{\mathbb{Z}}$ as in $\underline{g}_{\mathbb{Z}}$. Let r: $\underline{g}'_{\mathbb{Z}} \to \underline{g}_{\mathbb{Z}}$ be the canonical homomorphism of $\mathbb{Z}$ - modules. An element of ker r is a combination of H_α's which acts trivially on each X_β, because its image acts trivially on X_β. So r is a central extension and ker r is the centre $\underline{z}(\underline{g}'_{\mathbb{Z}})$ of $\underline{g}'_{\mathbb{Z}}$ because $\underline{z}(\underline{g}_{\mathbb{Z}}) = 0$.

2.15. PROPOSITION.

<u>The centre of</u> $\underline{g}'_{\mathbb{Z}}$ <u>is a direct sum of cyclic groups of prime order. Its order is</u>:

2 <u>for</u> $B_l (l \geq 2)$

2^{l-1} <u>for</u> $C_l (l \geq 2)$

4 <u>for</u> F_4

6 <u>for</u> G_2

1 <u>for other types (i.e. for types with one root length</u>).

PROOF.

We use the following lemma.

2.16. LEMMA.

<u>Relations</u> (2) <u>and</u> (3) <u>of</u> 2.13. <u>are equivalent, except for the case that</u> α,β,γ <u>are short roots in</u> G_2, <u>in which case</u> (3) <u>is obtained from</u> (2) <u>by multiplication with a factor</u> 2.

PROOF of LEMMA.

We know that (3) is a $N_{\beta\gamma}$-multiple of (2) (see 2.13.). If $|N_{\beta\gamma}| = 1$ then we are done. Let $|N_{\beta\gamma}|$ be larger than 1. Then $\beta-\gamma \in \Sigma$. As $\beta+\gamma \in \Sigma$ too, and $(\alpha,\alpha) = (\beta+\gamma, \beta+\gamma) \leq (\beta,\beta) \leq (\gamma,\gamma)$, we see from inspection of rank 2 root systems that β and γ are short roots in G_2, making an angle $2\pi/3$. In this case (3) states that $2H_\alpha + 2H_\beta + 2H_\gamma = 0$.

Now we proceed with the proof of the Proposition. If every H_α is expressable in the H_{α_i} (α_i simple) by means of the relations

(1), (3), then ker r = 0. Using the lemma we see that this is true if root lengths are equal. So we only have to worry about types B_l, C_l, F_4, G_2.

We use the description of Σ in terms of the ε_i (see [4], cf. 2.1), except in case G_2.

1). Let Σ be of type B_l, $\Sigma = \{\pm \varepsilon_i \pm \varepsilon_j, \pm \varepsilon_i\}$.

Relations (2) (or (3)) yield

(i) Relations involving only long roots.

(ii) Relations of the type $H_{\varepsilon_1} + H_{\varepsilon_2} + 2H_{-\varepsilon_1-\varepsilon_2} = 0$.

So after reduction mod 2 no interaction between long roots and short roots exists, i.e. every relation $\sum_{\alpha\in\Sigma} n_\alpha\{H_\alpha\} = 0$ implies a relation $\sum_{\alpha \text{ short}} n_\alpha\{H_\alpha\} = 0$, where $n_\alpha \in \mathbb{Z}$ and $\{H_\alpha\} = H_\alpha + 2\underline{g}'_{\mathbb{Z}}$.

Set

$$H = H_{\varepsilon_1+\varepsilon_2} + H_{\varepsilon_1-\varepsilon_2} + H_{-\varepsilon_1}.$$

As $\{H_{-\varepsilon_1}\} \neq 0$, we see that $\{H\} \neq 0$, hence $H \neq 0$. On the other hand $2H = (2H_{\varepsilon_1+\varepsilon_2} + H_{-\varepsilon_1} + H_{-\varepsilon_2}) + (2H_{\varepsilon_1-\varepsilon_2} + H_{-\varepsilon_1} + H_{\varepsilon_2}) = 0$.

Now we add relation $H = 0$ to relations (1), (2). Then every H_α is expressible in the H_β with β long, and hence in $H_{\varepsilon_1-\varepsilon_2}, \ldots, H_{\varepsilon_{l-1}-\varepsilon_l}, H_{\varepsilon_{l-1}+\varepsilon_l}$. This implies that we have got a full set of relations for $\underline{g}_{\mathbb{Z}}$ from those for $\underline{g}'_{\mathbb{Z}}$, in adding relation $H = 0$. We may conclude that H generates the centre of $\underline{g}'_{\mathbb{Z}}$, which is of order 2.

2). Let Σ be of type C_l. $\Sigma = \{\pm \varepsilon_i \pm \varepsilon_j, \pm 2\varepsilon_i\}$. Now relations (1), (2) yield

(i) Relations involving only short roots.

(ii) Relations of the type $H_{\varepsilon_1+\varepsilon_2} + H_{\varepsilon_1-\varepsilon_2} + 2H_{-2\varepsilon_1} = 0$.

Again there is no interaction between long roots and short roots after reduction mod 2. We see that the elements $H_i = H_{2\varepsilon_i} + H_{2\varepsilon_{i+1}} + H_{-\varepsilon_i-\varepsilon_{i+1}}$ $(1 \leq i \leq l-1)$, induce independent elements $\{H_i\}$ in $\mathfrak{g}'_{\mathbb{Z}}$ mod $2\mathfrak{g}'_{\mathbb{Z}}$.

And again $2H_i = 0$. After adding relations $H_i = 0$ to (1), (2) we can get rid of all H_β with β short, which proves as above that the H_i generate the centre.

3). Let Σ be of type F_4, $\Sigma = \{\pm \varepsilon_i \pm \varepsilon_j, \pm \varepsilon_i, \frac{\pm\varepsilon_1\pm\varepsilon_2\pm\varepsilon_3\pm\varepsilon_4}{2}\}$.

Set $\zeta = \dfrac{\varepsilon_1+\varepsilon_2+\varepsilon_3+\varepsilon_4}{2}$.

Relations (2) yield

(i) Relations involving only long roots.

(ii) Relations involving only short roots.

(iii) Relations of the type $H_{\varepsilon_1} + H_{\varepsilon_2} + 2H_{-\varepsilon_1-\varepsilon_2} = 0$.

Set $H_1 = H_{\varepsilon_1+\varepsilon_2} + H_{\varepsilon_1-\varepsilon_2} + H_{-\varepsilon_1}$.

$H_2 = H_{-\varepsilon_1-\varepsilon_2} + H_{-\varepsilon_3-\varepsilon_4} + H_\zeta$.

As in the case of B_l, we see that $H_i \neq 0$, $2H_i = 0$, $H_1 + H_2 \neq 0$. We want to show that adding relations $H_i = 0$ to relations (1), (2) yields a full set of relations for $\mathfrak{g}_{\mathbb{Z}}$. As in the case of B_l, it is sufficient to show that every H_α with α short is expressible in H_β's with β long. So we divide out these H_β's too, and we look what is left. One gets: $H_{-\varepsilon_1} = H_\zeta = 0$, $H_{\varepsilon_i} + H_{\varepsilon_j} = 0$, $H_{\pm\varepsilon_i} = 0$,

$0 = H_{-\varepsilon_i} + H_\zeta = H_{\zeta-\varepsilon_i}$ and so on.

We conclude that H_1 and H_2 span the centre.

4). Let Σ be of type G_2. Put $\alpha = \alpha_1$, $\beta = \alpha_1 + \alpha_2$, $\gamma = -\alpha-\beta$. Then $\Sigma = \{\pm\alpha, \pm\beta, \pm\gamma, \pm(\alpha-\beta), \pm(\beta-\gamma), \pm(\gamma-\alpha)\}$.

After dividing out relations (1), relations (3) yield:

(i) $H_{\alpha-\beta} + H_{\beta-\gamma} + H_{\gamma-\alpha} = 0$

(ii) $2H_{\alpha} + 2H_{\beta} + 2H_{\gamma} = 0$

(iii) Relations of the type $H_{\alpha} + H_{-\gamma} + 3H_{\gamma-\alpha} = 0$.

Set $H = H_{\gamma-\alpha} + H_{\beta-\alpha} + H_{\alpha}$.
After reduction mod 3 no interaction between long roots and short roots exists, so we may conclude as above that $2H \neq 0$.
After reduction mod 2 we see that $\{H_{\alpha}\}$, $\{H_{\beta}\}$, $\{H_{\gamma}\}$ are independent, so $\{3H\} = \{H_{\alpha} + H_{\beta} + H_{\gamma}\} \neq 0$ and hence $3H \neq 0$.
But $6H = 2(H_{\alpha} + H_{-\gamma} + 3H_{\gamma-\alpha}) + 2(H_{\alpha} + H_{-\beta} + 3H_{\beta-\alpha}) + (2H_{\alpha} + 2H_{\beta} + 2H_{\gamma}) = 0$. We conclude that H generates a cyclic group of order 6, hence a direct sum of two cyclic groups of prime order.
The fact that H generates the centre is checked as above.

2.17. COROLLARY.

(i) *If* Σ *is of type* F_4 *or* B_l $(l \geqslant 2)$, *then* $H_{\varepsilon_1+\varepsilon_2} + H_{\varepsilon_1-\varepsilon_2} + H_{-\varepsilon_1}$ *is an element of the centre that has a nonzero image in* $\mathfrak{g}'_{\mathbb{Z}}$ mod $2\mathfrak{g}'_{\mathbb{Z}}$.

(ii) *If* Σ *is of type* G_2, *then* $H_{\gamma-\alpha} + H_{\beta-\alpha} + H_{\alpha}$ *is an element of the centre that has a nonzero image in* $\mathfrak{g}'_{\mathbb{Z}}$ mod $3\mathfrak{g}'_{\mathbb{Z}}$, *and* $H'_{\beta-\gamma} + H'_{\alpha}$ *is an element of* $\mathfrak{g}'_{\mathbb{Z}}$ *that has a nonzero image in* $\mathfrak{g}'_{\mathbb{Z}}$ mod $2\mathfrak{g}'_{\mathbb{Z}}$.

§3. The action $\hat{\mathrm{Ad}}$. Structure of $\mathfrak{g}^*_{\mathbb{Z}}$, $\mathfrak{g}^*_R$.

In this section we describe the universal central extensions of $\mathfrak{g}_{\mathbb{Z}}$ and $\mathfrak{g}_k$, over $\mathbb{Z}$ and k respectively. Because of Proposition 1.3 (vi), knowledge of the universal central extension $\mathfrak{g}^*_{\mathbb{Z}} \to \mathfrak{g}_{\mathbb{Z}}$ of $\mathfrak{g}_{\mathbb{Z}}$ implies knowledge of that of $\mathfrak{g}_R$ for any ring R.

We can't apply this remark for type C_1 however, because of the fact that in this case $\underline{g}_{\mathbb{Z}}$ has no universal central extension (see Proposition 1.3, (ii), and Corollary 2.2, (ii)). But type C_1 is not very interesting, because its $\underline{g}_R$ is centrally closed as soon as the universal central extension $\underline{g}_R^* \to \underline{g}_R$ exists. (see 3.13).

3.1. Assume that we are in the situation of 2.1 and suppose that $\underline{g} = [\underline{g}, \underline{g}]$. G acts on $\underline{g}$ by the adjoint representation Ad. According to Proposition 1.3, (v), every automorphism Ad (x) of $\underline{g}$ induces a unique automorphism $\widehat{\mathrm{Ad}}$ (x) of $\underline{g}^*$. So we have a representation $\widehat{\mathrm{Ad}}$ of G(K) in $\underline{g}^*$. As $\underline{g}_k^* \to \underline{g}_k$ induces π: $\underline{g}^* \to \underline{g}$ (see Proposition 1.3,(vi)), we can take $\underline{g}^*$ to be defined over k in a natural way. Then π is defined over k.

3.2. REMARK.

There is exactly one k-structure of Lie algebras on $\underline{g}^*$ such that π is defined over k.

PROOF.

We make use of the central trick (1.2) again. Let $\underline{g}_k^*$ denote the k-structure from above, and $(\underline{g}^*)_k$ another one such that π is defined over k. Then $(\underline{g}^*)_k \subset [(\underline{g}^*)_k, (\underline{g}^*)_k] = [\underline{g}_k^*, \underline{g}_k^*] = \underline{g}_k^*$. It follows that $(\underline{g}^*)_k = \underline{g}_k^*$, because both are k-structures.

3.3. PROPOSITION.

$\widehat{\mathrm{Ad}}$: G $\to$ GL($\underline{g}^*$) is a homomorphism, defined over k (i.e. a k-morphism of algebraic groups). Its derivative d $\widehat{\mathrm{Ad}}$ (denoted $\widehat{\mathrm{ad}}$) is characterized by

$$\widehat{\mathrm{ad}} \circ \pi = \mathrm{ad},$$

where the right-hand side is the adjoint representation of $\underline{g}^*$ in $\underline{g}^*$.

PROOF.

We use the construction of $\underline{g}^*$, given in the proof of (ii), Proposition 1.3. Define the surjective homomorphism of K-modules $r: \underline{g} \otimes \underline{g} \to \underline{g}^*$ by $r(X \otimes Y) = \{X \otimes Y\}$. (Notations as in loc. cit). G acts on $\underline{g} \otimes \underline{g}$ by $Ad \otimes Ad$. As ker r (= N) is invariant under this action, an action Ad' of G on $\underline{g}^*$ is induced, with $Ad'(x)\ \{X \otimes Y\} = \{Ad(x)\ X \otimes Ad(x)\ Y\}$. Now Ad'(x) is a Lie algebra automorphism, satisfying $\pi \circ Ad'(x) = Ad(x) \circ \pi$, so $Ad' = \widehat{Ad}$. As $Ad \otimes Ad$ is a k-homomorphism, and ker r is defined over k, the representation $Ad' = \widehat{Ad}$ is a k-homomorphism.

As r is a homomorphism of G-modules, we have $r \circ (d(Ad \otimes Ad)\ (X)) = \widehat{ad}\ (X) \circ r$ for $X \in \underline{g}$.

So $\widehat{ad}\ (X)\{Y \otimes Z\} = \widehat{ad}(X)\ r(Y \otimes Z) = r(d(Ad \otimes Ad)(X)(Y \otimes Z)) = \{[X,Y] \otimes Z + Y \otimes [X,Z]\}$.

Hence $\widehat{ad}\ (\pi\{X \otimes X'\})\ \{[Y,Y'] \otimes [Z,Z']\} =$

$\{[[X,X'], [Y,Y']] \otimes [Z,Z'] + [Y,Y'] \otimes [[X,X'],[Z,Z']]\} =$

$= [\{X \otimes X'\}, \{[Y,Y'] \otimes [Z,Z']\}]$ because of Jacobi for $\{X \otimes X'\}$, $\{Y \otimes Y'\}$, $\{Z \otimes Z'\}$. Now the Proposition follows form the fact that $\underline{g} = [\underline{g}, \underline{g}]$.

3.4. The action $\widehat{Ad}$ makes $\underline{g}^*$ into a G-module. The maximal torus in $G_{\mathbb{C}}$, corresponding to the Cartan decomposition in $\underline{g}_{\mathbb{C}}$, gives rise to a k-split maximal torus T in G. The G-module $\underline{g}^*$ has a weight decomposition with respect to this torus. As, for $x \in G$, $\widehat{Ad}(x)$ is a Lie algebra automorphism, we see that the weight decomposition in $\underline{g}^*$ yields a structure of graded Lie algebras. This grading can give information about the structure of $\underline{g}^*$ as a Lie algebra. We are going to exploit an analogous grading for $\underline{g}^*_{\mathbb{Z}}$. We shall also use unipotent automorphisms of the type $\widehat{Ad}(x)$. (see 3.10).

3.5. THEOREM. (Structure of $g_{\mathbb{Z}}^*$). Assume that Σ is not of type C_1, $1 \geq 1$. The structure of the universal central extension $\pi: g_{\mathbb{Z}}^* \to g_{\mathbb{Z}}$ is as follows.

(i) As a $\mathbb{Z}$-module, $g_{\mathbb{Z}}^*$ is defined by the following generators and relations:

GENERATORS:

a) X_α^*, H_α^* $(\alpha \in \Sigma)$.

b) Z_γ^* (γ degenerate with respect to some n, which we denote n_γ). (See 2.6, (i)).

RELATIONS: (See 2.13).

(1) $H_\alpha^* + H_{-\alpha}^* = 0$ for $\alpha \in \Sigma$.

(3) $N_{\beta\gamma}\ H_\alpha^* + N_{\gamma\alpha}\ H_\beta^* + N_{\alpha\beta}\ H_\gamma^* = 0$ for $\alpha,\beta,\gamma \in \Sigma$ with $\alpha+\beta+\gamma=0$.

(4) $n_\gamma\ Z_\gamma^* = 0$.

(Relation (2) of 2.13 has been omitted).

(ii) The Lie algebra structure on $g_{\mathbb{Z}}^*$ is defined by:

$[X_\alpha^*, X_\beta^*] = N_{\alpha\beta}\ X_{\alpha+\beta}^*$ if $\alpha+\beta \in \Sigma$.

$[X_\alpha^*, X_{-\alpha}^*] = H_\alpha^*$.

$[X_\alpha^*, X_\beta^*] = \varepsilon(\alpha,\beta)\ Z_{\alpha+\beta}^*$ if α,β are independent and $\alpha+\beta$ is degenerate with respect to some n.

$[X_\alpha^*, X_\beta^*] = 0$ in other cases.

$[H_\alpha^*, X_\beta^*] = \langle\beta,\alpha\rangle\ X_\beta^*$.

$[H_\alpha^*, H_\beta^*] = 0$.

$[Z_\gamma^*, Y] = 0$ if $Y \in g_{\mathbb{Z}}^*$.

Here ε is a map $\Sigma \times \Sigma \to \{1,-1\}$ that satisfies $\varepsilon(\alpha,\beta)+\varepsilon(\beta,\alpha) = 0$ for $\alpha,\beta \in \Sigma$. (Every such map will do).

(iii) $\pi(X_\alpha^*) = X_\alpha$, $\pi(H_\alpha^*) = H_\alpha$, $\pi(Z_\gamma^*) = 0$.

PROOF. Let $\pi: \mathfrak{g}_{\mathbb{Z}}^* \to \mathfrak{g}_{\mathbb{Z}}$ be the universal central extension, as constructed in the proof of (ii), Proposition 1.3. We have to show that, given $\varepsilon: \Sigma \times \Sigma \to \{1,-1\}$, there are X_α^*, H_α^*, Z_γ^* as in the Theorem. There is a grading on $\mathfrak{g}_{\mathbb{Z}} \otimes \mathfrak{g}_{\mathbb{Z}}$ with values in Γ, corresponding to the weight decomposition with respect to Ad $\otimes$ Ad. There is also a grading on $\mathfrak{g}_{\mathbb{Z}}$, corresponding to the weight decomposition with respect to Ad. Let $r: \mathfrak{g}_{\mathbb{Z}} \otimes \mathfrak{g}_{\mathbb{Z}} \to \mathfrak{g}_{\mathbb{Z}}^*$ be defined by $r(X \otimes Y) = \{X \otimes Y\}$. (cf. proof of Proposition 3.3). Then ker r is homogeneous with respect to the grading on $\mathfrak{g}_{\mathbb{Z}} \otimes \mathfrak{g}_{\mathbb{Z}}$, and we may choose a grading on $\mathfrak{g}_{\mathbb{Z}}^*$, compatible with r. This grading is also compatible with π.

So we have a grading $\mathfrak{g}_{\mathbb{Z}}^* = \sum_\gamma (\mathfrak{g}_{\mathbb{Z}}^*)_\gamma$ satisfying

(5) $[(\mathfrak{g}_{\mathbb{Z}}^*)_\alpha, (\mathfrak{g}_{\mathbb{Z}}^*)_\beta] \subset (\mathfrak{g}_{\mathbb{Z}}^*)_{\alpha+\beta}$, which says that it is a grading, and

(6) $\pi(\mathfrak{g}_{\mathbb{Z}}^*)_\alpha \subset (\mathfrak{g}_{\mathbb{Z}})_\alpha$.

As $\mathfrak{g}_{\mathbb{Z}}$ is a free $\mathbb{Z}$ - module, we can choose a $\mathbb{Z}$ - linear section s of π. (We may even choose s compatible with the gradings, but we don't need that).

We see from (5), (6) and the central trick:

(7) $[s(\mathfrak{g}_{\mathbb{Z}})_\alpha , s(\mathfrak{g}_{\mathbb{Z}})_\beta] \subset (\mathfrak{g}_{\mathbb{Z}}^*)_{\alpha+\beta}$.

Using the central trick again, we get

(8) $\mathfrak{g}_{\mathbb{Z}}^* = [\mathfrak{g}_{\mathbb{Z}}^*, \mathfrak{g}_{\mathbb{Z}}^*] = [\sum_{\alpha\in\Sigma\cup(0)} s(\mathfrak{g}_{\mathbb{Z}})_\alpha, \sum_{\beta\in\Sigma\cup(0)} s(\mathfrak{g}_{\mathbb{Z}})_\beta] =$

$\sum_{\alpha,\beta\in\Sigma\cup(0)} [s(\mathfrak{g}_{\mathbb{Z}})_\alpha, s(\mathfrak{g}_{\mathbb{Z}})_\beta]$.

3.6. REMARK.

It is easily seen from the above, that there is exactly one grading on $\mathfrak{g}_{\mathbb{Z}}^*$ satisfying (6). After reduction mod p it yields the grading mentioned in (3.4), for the same reasons.

3.7. We are going to show now that the grading on $\underline{g}^*_{\mathbb{Z}}$ is a weight decomposition with respect to the analogue of the action $\hat{a}d$ (see (3.3)). Let $X \in (\underline{g}_{\mathbb{Z}})_\alpha$, $Y \in (\underline{g}_{\mathbb{Z}})_\beta$, $\alpha,\beta \in \Sigma \cup (0)$, $\delta \in \Sigma$.
Then we get from the Jacobi relation and the central trick:

$[sH_\delta, [sX, sY]] = [[sH_\delta, sX], sY] + [[sY, sH_\delta], sX] =$

$[<\alpha,\delta> sX, sY] + [-<\beta,\delta> sY, sX] = <\alpha+\beta,\delta> [sX, sY]$.

Combining with (7), (8) we see:

(9) $ad(sH_\delta)$ acts on $(\underline{g}^*_{\mathbb{Z}})_\gamma$ as scalar multiplication with $<\gamma,\delta>$.

(Compare with Proposition 3.3).

So $[s(\underline{g}_{\mathbb{Z}})_0, s(\underline{g}_{\mathbb{Z}})_0] = 0$.

As $(\underline{g}_{\mathbb{Z}})_\alpha$ is a $\mathbb{Z}$ - module of rank 1 for $\alpha \in \Sigma$, $[s(\underline{g}_{\mathbb{Z}})_\alpha, s(\underline{g}_{\mathbb{Z}})_\alpha] = 0$ for all $\alpha \in \Sigma \cup (0)$. We conclude that (8) can be sharpened to

(10) $$\underline{g}^*_{\mathbb{Z}} = \sum_{\substack{\alpha,\beta \in \Sigma \cup (0) \\ \alpha \neq \beta}} [s(\underline{g}_{\mathbb{Z}})_\alpha, s(\underline{g}_{\mathbb{Z}})_\beta].$$

As π is compatible with the gradings, ker π is homogeneous, i.e.

(11) $$\ker \pi = \sum_\gamma (\ker \pi)_\gamma.$$

Let $\gamma \in \Gamma$. If $\gamma = 0$, set $n_\gamma = 0$. If $\gamma \neq 0$, set $n_\gamma = \max\{n | \gamma \in n\Gamma\}$, or, equivalently, set

n_γ = g.c.d. of the $<\gamma,\delta>$, $\delta \in \Sigma$.

It is easily seen from lemma 2.6, (i), that this new definition of n_γ is an extension of the old one (see (i)).

We see from (9) that

(12) $n_\gamma(\ker \pi)_\gamma = 0$ $\qquad$ $(\ker \pi \subset \underline{z}(\underline{g}^*))$.

As we have excluded the types C_l, $l \geqslant 1$, we have $\Sigma \cap p\Gamma = \emptyset$ for every p (see 2.2). Hence

(13) $n_\gamma = 1$ for $\gamma \in \Sigma$.

We see from (10) that in $\underline{g}^*_{\mathbb{Z}}$ the only possible degrees are elements of $\Sigma \cup (0)$ and sums of independent roots. As ker π is contained in $\mathfrak{g}_{\mathbb{Z}}^*$, we conclude, using (12), (13) that

(14) $$\ker \pi = (\ker \pi)_0 + \sum_n \sum_{\substack{\gamma \text{ degenerate} \\ \text{sum with res-} \\ \text{pect to } n}} (\ker \pi)_\gamma .$$

Here we see how degenerate sums come into the picture.

Let $\alpha \in \Sigma$. As $(\ker \pi)_\alpha = 0$, we have an isomorphism

$\pi: (\mathfrak{g}_{\mathbb{Z}}^*)_\alpha \to (\mathfrak{g}_{\mathbb{Z}})_\alpha$.

Call it π_α.

(15) We choose X_α^* to be the inverse image of X_α under π_α.

(16) Define $H_\alpha^* = [X_\alpha^*, X_{-\alpha}^*]$ $(\alpha \in \Sigma)$, and define $Z_{\alpha,\beta}^* =$ $= \varepsilon(\alpha,\beta)[X_\alpha^*, X_\beta^*]$, if α,β are independent roots such that $\alpha+\beta$ is degenerate with respect to some n.

We have to show that $Z_{\alpha,\beta}^*$ depends only on $\alpha+\beta$. It is clear that $Z_{\alpha,\beta}^* = Z_{\beta,\alpha}^*$. (We require $\varepsilon(\alpha,\beta) + \varepsilon(\beta,\alpha) = 0$). Hence we consider the case that $\alpha,\beta,\gamma,\delta$ are distinct roots, while $\alpha+\beta = \gamma+\delta$ is degenerate with respect to some n.

In this case $n = 2$, $(\alpha,\beta) = (\gamma,\delta) = 0$, and we may suppose $(\alpha,\alpha) = (\gamma,\gamma) \leq (\beta,\beta) = (\delta,\delta)$. (See Lemma 2.6, (iii)).

Then $\langle\gamma,\beta\rangle \leq 1$, $\langle\delta,\beta\rangle \leq 1$, $\langle\gamma+\delta,\beta\rangle = \langle\alpha+\beta,\beta\rangle = 2$. So $\langle\gamma,\beta\rangle = 1$, and we have $\gamma-\beta \in \Sigma$.

Now suppose $\gamma-2\beta \in \Sigma$, to get a contradiction.

We have $\langle\gamma-2\beta,\beta\rangle = -3$, so Σ is of type G_2, β is short in Σ.

Then $(\alpha,\beta) = 0$ shows that $(\alpha,\alpha) > (\beta,\beta)$, a contradiction.

It follows that $N_{\gamma-\beta,\beta} = \pm 1$. For the same reasons $N_{\delta,\alpha-\delta} = -N_{\alpha-\delta,\delta} = \pm 1$ and $N_{\delta-\beta,\beta} = \pm 1$. Then $\beta+\delta \notin \Sigma$, since $\langle\delta,\beta\rangle = 1$ and $N_{\delta-\beta,\beta} = \pm 1$. Now we can compute the Jacobi relation for X_δ^*, $X_{\gamma-\beta}^*$, X_β^*, using the central trick:

$0 = [X_\delta^*, [X_{\gamma-\beta}^*, X_\beta^*]] + [X_\beta^*, [X_\delta^*, X_{\gamma-\beta}^*]] +$

$[X_{\gamma-\beta}^*, [X_\beta^*, X_\delta^*]] = N_{\gamma-\beta,\beta} [X_\delta^*, X_\gamma^*] +$

$N_{\delta,\alpha-\delta} [X_\beta^*, X_\alpha^*] + 0 = \pm Z_{\delta,\gamma}^* \pm Z_{\beta,\alpha}^*.$

As $n = 2$, or $n_{\alpha+\beta} = n_{\gamma+\delta} = 2$, it follows from (12) that $Z_{\beta,\alpha}^* = Z_{\delta,\gamma}^*$. Hence

(17) $Z_{\alpha+\beta}^* = \varepsilon(\alpha,\beta) [X_\alpha^*, X_\beta^*]$

is a good definition.

Next we have to prove that X_α^*, H_α^*, Z_γ^* behave as described in the Theorem. Part (iii) is obvious.

The relation $[X_\alpha^*, X_\beta^*] = N_{\alpha,\beta} X_{\alpha+\beta}^*$ follows from (5) and (15).

Relations $[X_\alpha^*, X_{-\alpha}^*] = H_\alpha^*$ and $[X_\alpha^*, X_\beta^*] = Z_\gamma^*$ follow from the definitions (16), (17).

For other cases $[X_\alpha^*, X_\beta^*] = 0$ because it is an element of $(\ker \pi)_{\alpha+\beta}$, which is the zero module (see (14)).

Relations $[Z_\gamma^*, Y] = 0$ are obvious, and the action of H_α^* is the same as that of sH_α, which is described in (9).

This proves (ii).

Using the central trick, we see from (10) that $\underline{g}_{\mathbb{Z}}^*$ is generated as a $\mathbb{Z}$ - module by the elements $[X,Y]$ where $X,Y \in \{X_\alpha^*, H_\alpha^* | \alpha \in \Sigma\}$.

Then we see from (ii) that $\underline{g}_{\mathbb{Z}}^*$ is generated as a $\mathbb{Z}$ - module by the elements X_α^*, H_α^*, Z_γ^*. We still have to prove now that (1), (3), (4) are defining relations.

It will be sufficient to look for defining relations of all components $(\underline{g}_{\mathbb{Z}}^*)_\beta$, because $\underline{g}_{\mathbb{Z}}^*$ is the direct sum of the $(\underline{g}_{\mathbb{Z}}^*)_\beta$.

First we prove that relations (1), (3), (4) are satisfied. Relation (4) is a special case of (12). Relation (1) is obvious. Relation (3) is the Jacobi relation for X_α^*, X_β^*, X_γ^*. (See (ii) and see 2.13).

3.8. PROOF CONTINUED.

We still have to prove that relation (1), (3), (4) are sufficient to define $\mathfrak{g}^*_{\mathbb{Z}}$. Consider the central extension $r: \mathfrak{g}'_{\mathbb{Z}} \to \mathfrak{g}_{\mathbb{Z}}$ of Lie algebras over $\mathbb{Z}$ (see 2.14).

There is a homomorphism $\tau: \mathfrak{g}^*_{\mathbb{Z}} \to \mathfrak{g}'_{\mathbb{Z}}$ such that $r \circ \tau = \pi$.

The central trick proves

$\tau(H^*_\alpha) = [\tau X^*_\alpha, \tau X^*_{-\alpha}] = [X_\alpha, X_{-\alpha}] = H_\alpha$.

So there can't be more relations between the H^*_α, then there are between the H_α in $\mathfrak{g}'_{\mathbb{Z}}$.

This proves:

(18) The subspace $(\mathfrak{g}^*_{\mathbb{Z}})_0$, generated by the H^*_α, has (1) and (3) as defining relations.

The other components of the grading of $\mathfrak{g}^*_{\mathbb{Z}}$ are $\mathbb{Z}$ - modules with one generator. If $\alpha \in \Sigma$, then $(\mathfrak{g}^*_{\mathbb{Z}})_\alpha$ is generated by X^*_α. It is a free $\mathbb{Z}$ - module, because $(\mathfrak{g}_{\mathbb{Z}})_\alpha$ is free. If δ is degenerate with respect to n, then $(\mathfrak{g}^*_{\mathbb{Z}})_\delta$ has Z^*_δ as generator, and $nZ^*_\delta = n_\delta Z^*_\delta = 0$ (see (12)). As n is prime (see lemma 2.6, (i)), $(\mathfrak{g}^*_{\mathbb{Z}})_\delta$ is either zero or n-cyclic.

(19) So if we prove that $Z^*_\delta \neq 0$, then all components of the grading satisfy description (i), which proves the Theorem.

3.9. REMARK.

It is possible to check that the $\mathbb{Z}$ - module with bracket-operation $\mathfrak{g}^*_{\mathbb{Z}}$, that is described in (i), (ii), is in fact a Lie algebra. This yields a central extension of $\mathfrak{g}_{\mathbb{Z}}$, that we can use in the same way as we used the extension $\mathfrak{g}'_{\mathbb{Z}} \to \mathfrak{g}_{\mathbb{Z}}$.

We won't pursue this line; we will exploit the action $\hat{\mathrm{Ad}}$ instead (see 3.1), which is a more instructive way.

3.10. Fixing δ, we take $p = n_\delta$, $k = \mathbb{F}_p$ and return to the notations of 2.1, 3.4.

The universal central extension $\pi: \underline{g}_k^* \to \underline{g}_k$ is obtained from $\underline{g}_{\mathbb{Z}}^* \to \underline{g}_{\mathbb{Z}}$ by reduction mod p. So we have in $\underline{g}_k^*$ the images of X_α^*, H_α^*, Z_δ^* ($\alpha \in \Sigma$, γ degenerate).

We denote them by X_α^*, H_α^*, Z_γ^*.

Now it is sufficient to prove that $Z_\delta^* \neq 0$ in $\underline{g}_k^*$. We are going to give this proof case by case, using the classification of degenerate sums.

<u>case 1</u>. $p = 2$, types B_l $(l > 4)$ and type F_4. We have a natural grading on $\underline{g}^*$(see 3.4 and 3.6).

As Z_δ^* generates $\underline{g}_\delta^*$, all we have to show is that δ has non-zero multiplicity in the G-module $\underline{g}^*$. For the types under consideration there is one orbit of degenerate sums (see 2.8 Table 1). Multiplicities are invariant under the action of the Weyl-group, so we may suppose $\delta = 2\varepsilon_2$. (Notations as in 2.1, 2.16).

Using the central trick and the fact that $p = 2$ we see (cf.[2], (4.5) (2))

$$\widehat{Ad}\,(x_{-\varepsilon_2}(1))\, Z_{2\varepsilon_2}^* = [\widehat{Ad}\,(x_{-\varepsilon_2}(1))\, X_{\varepsilon_1+\varepsilon_2}^*,\ \widehat{Ad}\,(x_{-\varepsilon_2}(1))\, X_{-\varepsilon_2+\varepsilon_2}^*]$$

$$= [X_{\varepsilon_1+\varepsilon_2}^* + X_{\varepsilon_1}^* + X_{\varepsilon_1-\varepsilon_2}^*,\ X_{-\varepsilon_1+\varepsilon_2}^* + X_{-\varepsilon_1}^* + X_{-\varepsilon_1-\varepsilon_2}^*] =$$

$Z_{2\varepsilon_2}^* + H_{\varepsilon_1+\varepsilon_2}^* + H_{-\varepsilon_1}^* + H_{\varepsilon_1-\varepsilon_2}^* + Z_{-2\varepsilon_2}^*$, which has non-zero component in $\underline{g}_0^*$. (See Corollary 2.17 and use the part of (i) that has been proved above). So $Z_{2\varepsilon_2}^*$ has non-zero image, which shows that $Z_{2\varepsilon_1}^*$ itself is non-zero.

<u>case 2</u>. $p = 2$, type B_4. We denote $\underline{g}_X$ the Lie algebra $\underline{g}$ of type X, and G_X the (simply connected) Chevalley group G of type X.

In $(\mathfrak{g}_{\mathbb{C}})_{F_4}$ there is a subalgebra generated by the weight components $(\mathfrak{g}_{\mathbb{C}})_{\pm\varepsilon_i}$, $(\mathfrak{g}_{\mathbb{C}})_{\pm\varepsilon_i\pm\varepsilon_j}$ $(i \neq j)$. This subalgebra is a semisimple algebra of type B_4 (see [14], § 5). The Chevalley basis in $(\mathfrak{g}_{\mathbb{C}})_{F_4}$ obviously induces a Chevalley basis in this subalgebra $(\mathfrak{g}_{\mathbb{C}})_{B_4}$. Hence there is an inclusion map $(\mathfrak{g}_{\mathbb{Z}})_{B_4} \to (\mathfrak{g}_{\mathbb{Z}})_{F_4}$, which induces a homomorphism of $\mathfrak{g}_{B_4}$ into $\mathfrak{g}_{F_4}$, sending $X_{\pm\varepsilon_i}$ to $X_{\pm\varepsilon_i}$ and $X_{\pm\varepsilon_i\pm\varepsilon_j}$ to $X_{\pm\varepsilon_i\pm\varepsilon_j}$. So there is a homomorphism $\mathfrak{g}^*_{B_4} \to \mathfrak{g}^*_{F_4}$ sending $Z^*_{\pm 2\varepsilon_i}$ to $Z^*_{\pm 2\varepsilon_i}$ and $Z^*_{\pm\varepsilon_1\pm\varepsilon_2\pm\varepsilon_3\pm\varepsilon_4}$ to $Z^*_{\pm\varepsilon_1\pm\varepsilon_2\pm\varepsilon_3\pm\varepsilon_4}$.

These Z^*_γ cover all possibilities (see 2.8, Table 1). So the image of Z^*_δ is non-zero, which proves that it is itself non-zero.

<u>case 3</u>. $p = 2$, type D_4.

Use the "trivial" homomorphism $\mathfrak{g}_{D_4} \to \mathfrak{g}_{B_4}$ that is analogous to the homomorphism $\mathfrak{g}_{B_4} \to \mathfrak{g}_{F_4}$.

<u>case 4</u>. $p = 2$, type B_3.

In this case we use a less trivial homomorphism.

3.11. DIGRESSION.

Let σ be a graph automorphism of G_{D_4} that has order 2. Say σ interchanges α_3 and α_4. The fixed point group $(G_{D_4})_\sigma$ of σ is an almost simple group of type B_3. This is easily seen from Theorem 8.2 in Steinberg [24], step (2) in the proof of this Theorem, Remark (b) following the proof.

The group $(G_{D_4})_\sigma$ has a maximal torus T_σ, consisting of fixed points in the torus $T = T_{D_4}$. So there is a homomorphism $G_{B_3} \to G_{D_4}$, mapping T_{B_3} onto T_σ, whose image is $(G_{D_4})_\sigma$. (See [7], Exposé 23, Théorème 1). We make it more explicit. Let V be a complex vector space of dimen-

sion 8, with a non-degenerate symmetric bilinear form B of maximal Witt index. Say $v_1, \ldots, v_4, v_{-1}, \ldots, v_{-4}$ is a basis of V such that $B(v_i, v_j) = \delta_{i,-j}$. (Kronecker δ).
In the Clifford algebra associated to B, the elements $v_i v_j$ span a Lie algebra of type D_4 (see [16], Theorem 7, p. 231).
The elements $v_i v_{-i} - v_{-j} v_j = [v_i v_j, v_{-j} v_{-i}]$ ($|i| \neq |j|$) span a Cartan subalgebra.
Let α be a root in Σ_{D_4}, say $\alpha = s_1 \varepsilon_i + s_2 \varepsilon_j$ where $s_k = \pm 1$, $i \neq j$. If $s_1 i < s_2 j$, then we put $X_\alpha = v_{s_1 i}\, v_{s_2 j}$. If $s_1 i > s_2 j$, then i has to be interchanged with j. We get a Chevalley basis this way.
The counterpart of σ in characteristic 0 interchanges v_4, $-v_{-4}$ and fixes the other v_i's. (So it maps $X_{\varepsilon_1+\varepsilon_4} = v_1 v_4$ to $-v_1 v_{-4} = v_{-4} v_1 = X_{\varepsilon_1-\varepsilon_4}$). Its fixed points in the Clifford algebra form an algebra that is generated by $v_0 = v_4 - v_{-4}$ and the v_i with $|i| < 4$. This is a Clifford algebra again, associated to the subspace V^1 of V generated by v_0, $v_{\pm 1}$, $v_{\pm 2}$, $v_{\pm 3}$ (see [9], 2.1, II. 1.4). Put $X_{\pm\varepsilon_i} = v_{\pm i} v_0$.
The elements $X_{\pm\varepsilon_i}$, $X_{\pm\varepsilon_i \pm \varepsilon_j}$, $i,j = 1,2,3$, $i \neq j$, generate a Lie algebra of type B_3, and yield a Chevalley basis again. The v_i generate a $\mathbb{Z}$ - form of the larger Clifford algebra. If we apply the construction of Chevalley groups from admissible lattices to the representations (by left multiplication) of $(\underline{g}_{\mathbb{Z}})_{B_3}$ and $(\underline{g}_{\mathbb{Z}})_{D_4}$ in this $\mathbb{Z}$ - form, then we get a Chevalley group G_{B_3} that is contained in a Chevalley group G_{D_4}. The inclusion map is given by

$$x_{\pm\varepsilon_i}(t) \mapsto x_{\pm\varepsilon_i+\varepsilon_4}(t)\; x_{\pm\varepsilon_i-\varepsilon_4}(t) \qquad (i < 4)$$

$$x_{\pm\varepsilon_i\pm\varepsilon_j}(t) \mapsto x_{\pm\varepsilon_i\pm\varepsilon_j}(t) \qquad (i,j < 4).$$

We get a homomorphism $\underline{g}_{B_3} \to \underline{g}_{D_4}$ given by

(21) $X_{\pm\varepsilon_i} \to X_{\pm\varepsilon_i+4} - X_{\pm\varepsilon_i-\varepsilon_4}$,

$X_{\pm\varepsilon_i\pm\varepsilon_j} \to X_{\pm\varepsilon_i\pm\varepsilon_j}$.

Note that the same can be done for all pairs B_l, D_{l+1} ($l \geq 2$).

3.12. PROOF THEOREM 3.5 (CONCLUDED).

We return to the proof of case 4.

Consider the homomorphism $\underline{g}_{B_3} \to \underline{g}_{D_4}$ that is described by (21) in 3.11. Note that it is easy to check directly that this is a homomorphism, because $p = 2$. The homomorphism $\underline{g}^*_{B_3} \to \underline{g}^*_{D_4}$ that is induced, sends $Z^*_{\pm\varepsilon_1\pm\varepsilon_2\pm\varepsilon_3}$ to $Z^*_{\pm\varepsilon_1\pm\varepsilon_2\pm\varepsilon_3+\varepsilon_4} + Z^*_{\pm\varepsilon_1\pm\varepsilon_2\pm\varepsilon_3-\varepsilon_4}$ and sends $Z^*_{\pm2\varepsilon_i}$ to $Z^*_{\pm2\varepsilon_i}$. Again, these Z^*_γ cover all posibilities.

<u>case 5</u>. $p = 2$, types $A_3 = D_3$ and D_l ($l > 4$).

Use the "trivial" homomorphism $\underline{g}_{D_l} \to \underline{g}_{B_l}$, cf. 3.10, case 3.

<u>case 6</u>. $p = 2$, type G_2.

In characteristic 2 there is a surjective homomorphism $\underline{g}_{A_3} \to \underline{g}_{G_2}$, having the centre of $\underline{g}_{A_3}$ as kernel.

It sends X_α to $X_{pr(\alpha)}$, where pr is the projection of the root system of type A_3 on a plane through a subsystem of type A_2. The image of this projection is a root system of type G_2. (Say $\Sigma_{A_3} = \{\varepsilon_i-\varepsilon_j \mid i \neq j,\ i,j \leq 4\}$ and project ε_1 on α_1, ε_2 on $\alpha_1+\alpha_2$, ε_3 on $-2\alpha_1-\alpha_2$, ε_4 on 0).

The existence of the corresponding homomorphism is easily checked. As $\underline{g}^*_{A_3} \to \underline{g}_{A_3} \to \underline{g}_{G_2}$ yields a central extension (see Proposition 1.3, (i)), there is a homomorphism $\mathfrak{g}^*_{G_2} \to \mathfrak{g}^*_{A_3}$. This homomorphism sends Z^*_δ to a nonzero element Z^*_γ, so we are done.

REMARKS.

1) In fact $\mathfrak{g}^*_{A_3} \cong \mathfrak{g}^*_{G_2}$. This is easily proved from the generalities in 1.3, using the fact that $\mathfrak{g}_{A_3} \to \mathfrak{g}_{G_2}$ is a central extension. Then one can see again that $\mathfrak{g}^*_\delta \neq 0$, from the dimensions of $\mathfrak{g}^*_{G_2}$ and $\mathfrak{g}^*_{A_3}$.

2) Case 6 can also be handled like case 4. There is a homomorphism $\mathfrak{g}_{G_2} \to \mathfrak{g}_{D_4}$, reflecting the fact that the graph automorphism of order 3 in Spin_8 has a fixed point group of type G_2. (see [24], § 8 and [22], p. 176, (c)).

Note that this homomorphism $\mathfrak{g}_{G_2} \to \mathfrak{g}_{D_4}$ also exists in other characteristics, contrary to the homomorphism $\mathfrak{g}_{A_3} \to \mathfrak{g}_{G_2}$.

3) Finally, case 6 can also be handled like case 1.

<u>case 7</u>. $p = 3$, type G_2.

We proceed as in the case of $p = 2$, type F_4, using the same notations for the roots as in 2.16, case 4.

It is sufficient to show that $Z^*_{3\alpha} \neq 0$. Using the central trick we see

$\widehat{\mathrm{Ad}}\,(x_{-\alpha}(1))\, Z^*_{3\alpha} =$

$[\widehat{\mathrm{Ad}}\,(x_{-\alpha}(1))\, X^*_{\alpha-\gamma},\ \widehat{\mathrm{Ad}}\,(x_{-\alpha}(1))\, X^*_{\alpha-\beta}] =$

$[X^*_{\alpha-\gamma} \pm X^*_{-\gamma} \pm X^*_\beta \pm X^*_{\beta-\alpha},\ X^*_{\alpha-\beta} \pm X^*_{-\beta} \pm X^*_\gamma \pm X^*_{\gamma-\alpha}]$.

So its component H^* in $\mathfrak{g}^*_0$ is

$H^*_{\alpha-\gamma} + c_1 H^*_\gamma + c_2 H^*_\beta + c_3 H^*_{\beta-\alpha},\ c_i \in \mathbb{F}_3$.

As a special case of relation (3) (see 3.5), we get

$0 = N_{\beta,-\gamma} H^*_{\gamma-\beta} + N_{-\gamma,\gamma-\beta} H^*_{\beta} + N_{\gamma-\beta,\beta} H^*_{-\gamma} = \pm H^*_{\beta} \pm H^*_{\gamma}$.

(One also can use Lemma 2.16).

In the same way $\pm H^*_\alpha \pm H^*_\gamma = 0$. So $H^* = H^*_{\alpha-\gamma} + c_3 H^*_{\beta-\alpha} + c_4 H^*_\alpha$. ($c_4 \in \mathbb{F}_3$).

Suppose $H^* = 0$. Then $H_{\alpha-\gamma} + c_3 H_{\beta-\alpha} + c_4 H_\alpha = 0$ in $\underline{g}$, so $c_3 = c_4 = -1$.

(In fact these relations hold without the assumption). But

$H^*_{\alpha-\gamma} - H^*_{\beta-\alpha} - H^*_\alpha \neq 0$ (see Corollary 2.17, (ii)). So $Z^*_{3\alpha} \neq 0$.

case 8. $p = 3$, type A_2.

Use the "trivial" homomorphism $\underline{g}_{A_2} \to \underline{g}_{G_2}$, cf. 3.10, case 3.

It is seen from 2.8, Table 1, that we have dealt with all possibilities for δ.

3.13. PROPOSITION. Let Σ be of type C_l, $l \geqslant 1$. Let R be a ring. If $[\underline{g}_R, \underline{g}_R] = \underline{g}_R$, then $\underline{g}_R$ is centrally closed.

PROOF.

The finitely generated $\mathbb{Z}$ - module $\underline{g}_{\mathbb{Z}} / [\underline{g}_{\mathbb{Z}}, \underline{g}_{\mathbb{Z}}]$ has 2-torsion, because all $2X_\alpha = [H_\alpha, X_\alpha]$ are in $[\underline{g}_{\mathbb{Z}}, \underline{g}_{\mathbb{Z}}]$, while some X_α are not. (see 2.2, Corollary). So if $\underline{g}_R = [\underline{g}_R, \underline{g}_R]$, or, equivalently, if $(\underline{g}_{\mathbb{Z}} / [\underline{g}_{\mathbb{Z}}, \underline{g}_{\mathbb{Z}}]) \otimes_{\mathbb{Z}} R = 0$, then

(0) $\frac{1}{2} \in R$.

Now we proceed as in the proof of Theorem 3.5 with $\mathbb{Z}$ replaced by R. Starting from the grading on $\underline{g}_R \otimes_R \underline{g}_R$ we get a grading on $\underline{g}^*_R$. Again we choose a section s of π, and we get the formulas

(10) $\underline{g}^*_R = \sum_{\substack{\alpha,\beta \in \Sigma \cup (0) \\ \alpha \neq \beta}} [s(\underline{g}_R)_\alpha, s(\underline{g}_R)_\beta]$ (see 3.7, relation (10)),

(11) $\ker \pi = \sum_\gamma (\ker \pi)_\gamma$ (see 3.7, relation (11)),

(12) $n_\gamma (\ker \pi)_\gamma = 0$ (see 3.7, relation (12)).

Now n_γ is either 1 or 2 for $\gamma \neq 0$, $(\mathfrak{g}_R^*)_\gamma \neq 0$. (Use (10) and see 2.8, Table 1). So relations (0), (12) imply $(\ker \pi)_\gamma = 0$ for $\gamma \neq 0$. Relations (1), (3) of 2.13 or 3.5 hold again, for the same reasons as in 3.7. (Define H_α^* in the same way).

There is a canonical surjection $\mathfrak{g}'_{\mathbb{Z}} \otimes_{\mathbb{Z}} R \to \mathfrak{g}_R^*$ (see 2.14). As the centre of $\mathfrak{g}'_{\mathbb{Z}}$ is a group of order 2^{l-1} (see 2.15), $\mathfrak{g}'_{\mathbb{Z}} \otimes R$ is canonically isomorphic to $\mathfrak{g}_R$ (Use (0)). So $\mathfrak{g}'_{\mathbb{Z}} \otimes R \to \mathfrak{g}_R^* \to \mathfrak{g}_R$ is an isomorphism, and π is an isomorphism.

3.14. COROLLARY.

Let $\Sigma \cap p\Gamma = \emptyset$ (see (2.2)).

(i) $\mathfrak{g}_k$ is centrally closed if and only if there is no degenerate sum.

(ii) For each degenerate sum, its multiplicity in $\mathfrak{g}^*$ is 1.

(iii) Every non-zero weight of $\ker \pi$ is degenerate.

(iv) a. If root lengths are equal, then $(\ker \pi)_0 = 0$.

b. If Σ is of type F_4 and $p = 2$, then $\dim (\ker \pi)_0 = 2$.

c. If Σ is of type B_l and $p = 2$, then $\dim (\ker \pi)_0 = 1$.

d. If Σ is of type G_2 and $p = 2$ or $p = 3$, then $\dim (\ker \pi)_0 = 1$.

(Note that cases a, b, c, d cover all possibilities for the occurrence of degenerate sums).

PROOF. See 2.15, 3.5, 3.13 and 2.8, Table 1.

3.15. COROLLARY.

$\mathfrak{g}_{\mathbb{C}}$ is centrally closed for all types.

3.16. COROLLARY.

Let $\Sigma \cap p\Gamma = \emptyset$. Then $6 \ker(\pi : \mathfrak{g}_{\mathbb{Z}}^* \to \mathfrak{g}_{\mathbb{Z}}) = 0$.

PROOF. See 2.6, 2.15, 3.5.

3.17. Put $l^* = \dim \underline{g}_0^*$ and $d^* = \dim \underline{g}^*$.

We get the following list:

$p = 2$	type A_3	$l^* = 3$	$d^* = 21$
	type B_3	$l^* = 4$	$d^* = 36$
	type B_4	$l^* = 5$	$d^* = 61$
	type B_l $(l > 4)$	$l^* = l+1$	$d^* = 2l^{*2}-l^*$
	type D_4	$l^* = 4$	$d^* = 52$
	type D_l $(l > 4)$	$l^* = l$	$d^* = 2l^2+l$
	type F_4	$l^* = 6$	$d^* = 78$
	type G_2	$l^* = 3$	$d^* = 21$
$p = 3$	type A_2	$l^* = 2$	$d^* = 14$
	type G_2	$l^* = 3$	$d^* = 21$

Put $d = \dim \underline{g}$. Then we have the following partial list: (note the resemblance)

type B_3	$l = 3$	$d = 21$
type B_4	$l = 4$	$d = 36$
type D_l $(l > 5)$	$l = l$	$d = 2l^2-l$
type F_4	$l = 4$	$d = 52$
type B_l $(l > 4)$	$l = l$	$d = 2l^2+l$
type E_6	$l = 6$	$d = 78$
type G_2	$l = 2$	$d = 14$

Question: Why is the pair 5,61 not present in the second list?

§4. Admissible lattices and the category $\mathcal{L}_V$. Σ-connected components.

In the sequel we shall make a frequent use of the construction of Chevalley groups from admissible lattices. So it will be convenient to list in this section some properties and notations.

4.1. NOTATIONS AND DEFINITIONS.

Let p, k, K, $\mathfrak{g}$, G, T,... be as above (see 2.1, 3.4). Let $\mathcal{U}$ be the universal enveloping algebra of $\mathfrak{g}_{\mathbb{C}}$ over $\mathbb{C}$. The $\mathbb{Z}$ - form generated by the $X_\alpha^n/n!$ ($\alpha \in \Sigma$, $n \geq 0$) is denoted $\mathcal{U}_{\mathbb{Z}}$.

Let ρ be a faithful representation of $\mathfrak{g}_{\mathbb{C}}$ in a complex vector space V. (All dimensions are finite).

The canonical extension of ρ to $\mathcal{U}$ will also be denoted ρ.

A lattice in V is a $\mathbb{Z}$ - form of V, an admissible lattice M in V is a lattice that is invariant under $\rho(\mathcal{U}_{\mathbb{Z}})$. (See [8]). If V is irreducible, then a standard lattice in V is a lattice $\rho(\mathcal{U}_{\mathbb{Z}})v$, where v is a highest weight vector (see [2], Proposition 2.4).

Let M be an admissible lattice.

The K-module which has $\mathbb{F}_p$ -structure $M \otimes_{\mathbb{Z}} \mathbb{F}_p$ is denoted L_M.

So L_M is obtained by reduction mod p. The action of $\mathcal{U}_{\mathbb{Z}}$ on L_M and the representation of G in L_M are both denoted ρ_M. So

(1) $\rho_M(x_\alpha(t)) = \sum_{n \geq 0} t^n \rho_M(X_\alpha^n/n!)$ $(\alpha \in \Sigma,\ t \in K)$.

(see [2], 3.1). With the notations of loc. cit. one may choose a representation π such that $\Gamma_\pi = \Gamma_{sc}$. Then $G = G_{\pi,K}$ and its representation in L_M is $\lambda_{\rho,\pi}$.

REMARK 1. The action ρ_M of G on L_M is defined over $\mathbb{F}_p$. (See [2], 3.3 (2)).

If M' is another admissible lattice in V, such that $M \subset M'$, then this inclusion induces a homomorphism of K-modules $L_M \to L_{M'}$, defined over $\mathbb{F}_p$. It is a homomorphism of $\mathcal{U}_{\mathbb{Z}}$-modules too, so it is a homomorphism of G-modules. We denote the quotient $L_{M'/M}$. This notation is justified by the right exactness of the tensor product, which yields

(2) $L_{M'/M} \cong (M'/M) \otimes_{\mathbb{Z}} K.$

(So $L_{M'/M}$ is also obtained by reduction mod p).

Note that M'/M is a $\mathcal{U}_{\mathbb{Z}}$-module, and that the action $\rho_{M'/M}$ of G on $L_{M'/M}$ is related to the action $\rho_{M'/M}$ of $\mathcal{U}_{\mathbb{Z}}$ on $L_{M'/M}$ by the formula

(3) $\rho_{M'/M}(x_\alpha(t)) = \sum_{n\geq 0} t^n \rho_{M'/M}(X_\alpha^n/n!).$

REMARK 2.

The action $\rho_{M'/M}$ of G on $L_{M'/M}$ is defined over $\mathbb{F}_p$, because both $\rho_{M'}$ and $L_M \to L_{M'}$ are defined over $\mathbb{F}_p$.

(4) An element of $L_{M'/M}$, corresponding to $x \in M'$ will be denoted $\{x\}_{M'/M}$, or $\{x\}$.

We shall usually denote $\rho_{M'/M}(X_\alpha^n/n!)\ \{x\}$ as $X_\alpha^n/n! \cdot \{x\}$. Analogous conventions hold for L_M.

Let V be fixed.

The category of G-modules of type $L_{M'/M}$ with morphisms induced by inclusions of lattices is denoted $\mathcal{L}_V$.

(So a morphism $L_{M_1'/M_1} \to L_{M_2'/M_2}$ sends $\{x\}_{M_1'/M_1}$ to $\{x\}_{M_2'/M_2}$, where $M_1' \subset M_2'$, $M_1 \subset pM_2' + M_2$).

4.2. LEMMA.

<u>Let</u> $\rho: \mathfrak{g}_{\mathbb{C}} \to V$, $\rho': \mathfrak{g}_{\mathbb{C}} \to V'$ <u>be complex representations. Let</u> $\rho \otimes \rho'$ <u>be the tensor representation in</u> $V \otimes V'$. <u>Then</u>

$$(\rho \otimes \rho')\ (X_\alpha^n/n!)\ a \otimes b = \sum_{i=0}^{n} \binom{n}{i} \rho(X_\alpha^i/i!)\ a \otimes \rho'(X_\alpha^{n-i}/(n-i)!)b.$$

$(\alpha \in \Sigma,\ n \geq 0,\ a \in V,\ b \in V').$

PROOF.

This lemma is an easy consequence of the definition $(\rho \otimes \rho')Y = \rho(Y) \otimes 1 + 1 \otimes \rho'(Y)$. (cf. [22], Lemma 7).

4.3. LEMMA.

If M, M' *are admissible lattices in* V, V' *respectively, then* $M \otimes M'$ *is an admissible lattice in* $V \otimes V'$, *and* $M \oplus M'$ *is an admissable lattice in* $V \oplus V'$.

4.4. LEMMA.

Let M', M, V, $L_{M'/M}$ *be as in* 4.1. *Let* A *be a linear subspace of* $L_{M'/M}$. *Then* A *is* $\mathcal{U}_{\mathbb{Z}}$ *- invariant if and only if* A *is* G-*invariant*.

PROOF.

Let A be $\mathcal{U}_{\mathbb{Z}}$ - invariant. Then A is G-invariant because of 4.1, formula (3). Conversely, let A be G-invariant. As T is K-split, we have $A = \sum_{\gamma} A_{\gamma}$.

If $a \in A_{\gamma}$, $\alpha \in \Sigma$, then $\sum_{n \geq 0} (X_{\alpha}^{n}/n!) \cdot a \in A$. (Use 4.1, formula (3) again). Taking homogeneous parts, we see $(X_{\alpha}^{n}/n!) \cdot A \subset A$.

4.5. LEMMA.

(i) *If* A *is a* G-*submodule of* $L_{M'/M}$, *defined over* $\mathbb{F}_p$, *then the inclusion map* $A \to L_{M'/M}$ *is a morphism in the category* $\mathcal{L}_V$.

(ii) *If* ϕ *is a morphism in* $\mathcal{L}_V$, *then its cokernel and its kernel are in* $\mathcal{L}_V$.

PROOF.

(i) Let $r\colon M' \to L_{M'/M}$ be the canonical map. Then A is spanned by $r(r^{-1}(A))$. We have $M \subset r^{-1}(A) \subset M'$, so $r^{-1}(A)$ is a lattice. It is an admissible lattice because of lemma 4.4. Now we have the injection of $\mathbb{F}_p$-modules $r: r^{-1}(A)/\ker r \to A$, that induces an isomorphism $L_{r^{-1}(A)/\ker r} \to A$.

The map $L_{r^{-1}(A)/\ker r} \to L_{M'/M}$ is in $\mathcal{L}_V$.

(ii) From (i) it is clear that kernels are in $\mathcal{L}_V$. The cokernel of $L_{M_1'/M_1} \to L_{M_2'/M_2}$ is $L_{M_2'/M_2+M_1'}$.

4.6. NOTATION. Let A be a G-submodule of L_M, defined over $\mathbb{F}_p$. Then $\{v \in \frac{1}{p} M \mid \{pv\}_M \in A\}$ is denoted M_A.

LEMMA.

Let A, M be as above.

(i) M_A is an admissible lattice containing M, such that $A = \ker(L_M \to L_{M_A})$.

(ii) If M' is another admissible lattice, containing M, such that $A = \ker(L_M \to L_{M'})$, then $M \subset M_A \subset M'$.

PROOF.

(i) As $M \subset M_A$, M_A is a lattice. It is obvious that it is an admissible one. In order to prove that $A = \ker(L_M \to L_{M_A})$ we first note that both sides are defined over $\mathbb{F}_p$. So both sides are spanned by elements $\{v\}_M$. Now

$\{v\}_M \in A \Leftrightarrow \frac{1}{p} v \in M_A \Leftrightarrow \{v\}_M \in \ker(L_M \to L_{M_A})$.

(ii) If $v \in M_A$, then $\{pv\}_M \in A$, so $\{pv\}_{M'} = 0$. If follows that $pv \in pM'$, whence $v \in M'$.

4.7. REMARK.

If V is the direct sum of two proper G-submodules V_1, V_2, then there is a natural embedding $\mathcal{L}_{V_1} \oplus \mathcal{L}_{V_2} \to \mathcal{L}_V$. (The notion of a direct sum of two additive categories is obvious). But this embedding is not always an isomorphism. (There is no "complete reducibility" over $\mathbb{Z}$.)

EXAMPLE.

Let Σ be a root system such that degenerate sums (with respect to p) exist. In $\mathfrak{g}_{\mathbb{C}}$ the lattice $\mathfrak{g}_{\mathbb{Z}}$ is admissible (see [2], Proposition 2.6). So $\mathfrak{g}_{\mathbb{Z}} \otimes \mathfrak{g}_{\mathbb{Z}}$ is admissible in $\mathfrak{g}_{\mathbb{C}} \otimes \mathfrak{g}_{\mathbb{C}}$ (see 4.3).

There are homomorphisms of $\mathcal{U}_{\mathbb{Z}}$-modules

ϕ: $\mathfrak{g}_{\mathbb{C}} \otimes \mathfrak{g}_{\mathbb{C}} \to \mathfrak{g}_{\mathbb{C}}$, defined by $\phi(X \otimes Y) = [X,Y]$,

ψ: $\mathfrak{g}_{\mathbb{Z}} \otimes \mathfrak{g}_{\mathbb{Z}} \to \mathfrak{g}^*$, (see proof of Proposition 1.3, (ii)),

χ: $\mathfrak{g}_{\mathbb{Z}} \to \mathfrak{g}$,

π: $\mathfrak{g}^* \to \mathfrak{g}$.

Note that $\mathfrak{g}^* = L_{\mathfrak{g}_{\mathbb{Z}} \otimes \mathfrak{g}_{\mathbb{Z}}/\ker \psi}$.

Set $N = \ker \phi$.

Then $\mathfrak{g}_{\mathbb{C}} \otimes \mathfrak{g}_{\mathbb{C}} \cong N \oplus \mathfrak{g}_{\mathbb{C}}$, because of complete reducibility over $\mathbb{C}$.
But it is not possible to decompose $\mathfrak{g}_{\mathbb{Z}} \otimes \mathfrak{g}_{\mathbb{Z}}$ in the same way. Suppose it were: Say $\mathfrak{g}_{\mathbb{Z}} \otimes \mathfrak{g}_{\mathbb{Z}} = M_1 \oplus M_2$, where $M_1 \subset N$, both M_i are $\mathcal{U}_{\mathbb{Z}}$-modules. Then $(\pi \circ \psi)M_1 = (\chi \circ \phi)M_1 = 0$. So $\mathfrak{g}^*_{\alpha}$ $(\alpha \in \Sigma)$ must be spanned by ψM_2.

The span of ψM_2 is a G-module (see Lemma 4.4). So it is invariant under $\widehat{ad}$ (see Proposition 3.3). Then it is clear from the description of $\mathfrak{g}^*$ that ψM_2 spans $\mathfrak{g}^*$ and not only the $\mathfrak{g}^*_{\alpha}$ $(\alpha \in \Sigma)$. But this is nonsense, because M_2 is an abelian group of rank equal to dim $\mathfrak{g}$, which is less then dim $\mathfrak{g}^*$.

It is easy to see from this example (and many others), that there may be "indecomposable" lattices in decomposable $\mathfrak{g}_{\mathbb{C}}$ - modules. (Definitions as below).

4.8. DEFINITION.

A G-module is called _indecomposable_ if it is not the direct sum of two non-trivial G-submodules.

4.9. LEMMA.

Every (finite dimensional) G-module L _is the direct sum of indecomposable submodules_ L_i.

4.10. DEFINITION.

The L_i in Lemma 4.9 are called the _indecomposable components_ of L.

REMARK. There is some "abuse of language" here: The L_i are not unique, but there is a Krull-Schmidt-Theorem (see [10], (14.5)).

4.11. LEMMA.

Let (ρ,V) _be an irreducible representation of_ $\mathfrak{g}_{\mathbb{C}}$. _Let_ M_{st} _be a standard lattice in_ V (_see_ 4.1).

Then for every admissible lattice $M \subset M_{st}$, _the_ G-_module_ $L_{M_{st}/M}$ _is indecomposable_.

PROOF.

Let v be the highest weight vector that generates M_{st}. Let λ be the highest weight. Suppose $L = L_{M_{st}/M}$ has decomposition $A \oplus B$. As λ has multiplicity 1 in V, it has at most multiplicity 1 in L, so $\{v\} \in A_\lambda$ or $\{v\} \in B_\lambda$. As $\{v\}$ generates L, we have A = L or B = L.

4.12. DEFINITIONS.

Let $\alpha,\beta \in \Gamma$ (see (2.1)). Then α,β are called _directly Σ-connected_ if there is $\gamma \in \Sigma$, $n \in \mathbb{Z}$, such that $\alpha-\beta = n\gamma$.

Now let Δ be a subset of Γ. We say that α,β are _directly Σ-connected in Δ_ if they are in Δ and are directly Σ-connected. (So γ need not be in Δ.) The transitive closure of the relation "directly Σ-connected in Δ" is called "Σ-connected in Δ". Equivalently, α,β are called <u>Σ-connected in Δ</u> if there is a sequence $\zeta_1,\ldots,\zeta_n$ of elements of Δ, such that

(i) $\zeta_1 = \alpha$, $\zeta_n = \beta$

(ii) For $1 \leq i < n$ the elements ζ_i, ζ_{i+1} are directly Σ-connected in Δ.

The equivalence classes with respect to the relation "Σ-connected in Δ" are called the <u>Σ-connected components</u> of Δ.

4.13. LEMMA.

Let L be a G-module, Δ its set of weights.

Let $\Delta_1,\ldots,\Delta_n$ be the Σ-connected components of Δ.

Put $L_i = \sum_{\lambda\in\Delta_i} L_\lambda$.

Then

(i) *Each indecomposable G-submodule of L is contained in some L_i.*

(ii) *The L_i are G-submodules.*

(iii) $L = \bigoplus_i L_i$.

PROOF.

(ii) If $\lambda \in \Delta_i$, $v \in L_\lambda$, $\alpha \in \Sigma$, then $x_\alpha(t).v \in \sum_{j\geq 0} L_{\lambda+j\alpha}$ (see [2], Lemma 5.2).

As the $\lambda+j\alpha$ are directly Σ-connected to α, we see that $x_\alpha(t).L_i \subset L_i$, so $G.L_i \subset L_i$.

(iii) is obvious.

(i) Let L' be an indecomposable G-submodule of L. As $L' = \bigoplus_\lambda L'_\lambda$, we have $L' = \bigoplus_i (L' \cap L_i)$.

But L' is indecomposable, so there is only one non-trivial term in $\bigoplus_i (L' \cap L_i)$.

4.14. NOTATION.

Suppose $L_{M'/M}$ has a composition series $L_{M'/M} = L_1 \supset L_2 \supset \dots$ $\dots \supset L_{k+1} = (0)$ whose elements are defined over $\mathbb{F}_p$, hence are in $\mathcal{L}_V$ (see 4.5). Then we denote this composition series by $v_1/v_2/\dots/v_k$, where $v_i \in M'$, such that $\{v_i\}_{M'/M} \in L_i$, $\{v_i\}_{M'/M} \notin L_{i+1}$. (So L_i is generated by $\{v_i\}, \{v_{i+1}\},\dots, \{v_k\}$).

REMARK.

Such a composition series always exists (see [27], Corollary 3F).

§5. The G-module ker π.

In this section we will study the restriction of $\hat{Ad}$ to ker π.

5.1. There is a Frobenius endomorphism Fr of G, sending $x_\alpha(t)$ to $x_\alpha(t^p)$ (see [7], Exposé 23, Théorème 1). Let $\delta^1,\dots, \delta^r$ be the fundamental weights such that $p\delta^i$ is a degenerate sum (see Lemma 2.9, (i)). Let (ρ_i, V_i) be the irreducible representation of $\mathfrak{g}_{\mathbb{C}}$ with highest weight δ^i, M_i a standard lattice in V_i.

5.2. PROPOSITION.

Assume $\Sigma \cap p\Gamma = \emptyset$. Let G act on ker π by the action $\hat{Ad}$, and let R be an indecomposable component of ker π (see 3.1, 3.3). Then there is a fundamental weight δ^i as above, such that

(i) All non-zero weights of R are in the orbit of $p\delta^i$,

(ii) $R_0 \neq 0$ if and only if $\delta^i \in \Sigma$,

(iii) The representation of G in R is $\mathbb{F}_p$-isomorphic to $(\rho_i)_{M_i} \circ \mathrm{Fr}$,

(iv) R is irreducible, except for the case that $p = 2$ and δ^i is a short root in B_1 or G_2. Then R_0 is a 1-dimensional G-submodule, and R/R_0 is irreducible.

PROOF.

Let Δ be the set of weights of ker π.

We know that each nonzero element of Δ is a degenerate sum (see Corollary 3.14, (iii)).

(1) We claim that the Σ-connected components Δ_i of Δ (see 4.12), are sets of the following types:

type a. An orbit of $p\delta^i$, where δ^i is a fundamental weight, $\delta^i \notin \Sigma$.

type b. The union of (0) and the orbit of $p\delta^i$, where δ^i is a fundamental weight, $\delta^i \in \Sigma$.

Proof of (1).

First we note that for every $\alpha \in \Sigma$, $\gamma \in \Delta$, the weights γ and $w_\alpha \gamma = \gamma - \langle\gamma,\alpha\rangle\alpha$ are directly Σ-connected. So the Σ-connected components are invariant under the action of W.

If δ^i is a root, then 0, $p\delta^i$ are directly Σ-connected. We see that sets of type a or b are Σ-connected. Their union is Δ, so we have to prove now:

If $\alpha,\beta \in \Delta (\alpha \neq \beta)$ are directly Σ-connected, then they are contained in the same set of type a or b. Say $\alpha-\beta = n\gamma$, $\gamma \in \Sigma$, $n \in \mathbb{Z}$. If α or β is zero, then it is easy. So suppose both are degenerate sums. The reflection w_γ leaves invariant the line $L_{\alpha,\beta}$ through α,β. It interchanges vectors of equal length. So if α,β have equal lengths, then $w_\gamma\alpha = \beta$ and we are done.

If α,β have different lengths, then Σ is of type B_3. (See 2.11 and use 2.2 to show that $\Sigma \cap p\Gamma = \emptyset$ excludes type C_1).

Hence we have $\Sigma = \{\pm\varepsilon_i, \pm\varepsilon_i \pm\varepsilon_j\}$, $\Delta = \{0, \pm 2\varepsilon_i, \pm\varepsilon_1\pm\varepsilon_2\pm\varepsilon_3\}$. Put $\Delta_a = \{0, \pm 2\varepsilon_i\}$, $\Delta_b = \{\pm\varepsilon_1\pm\varepsilon_2\pm\varepsilon_3\}$.

The sets Δ_a, Δ_b are of type a,b respectively. As α,β have different lengths, one is in Δ_a, one is in Δ_b. We see that α,β differ in all coordinates with respect to the ε_i. This yields a contradiction, as α,β are directly Σ-connected.

We have proved claim (1) now.

(2) Next we claim that every indecomposable G-submodule of $\ker \pi$ is of type $L_i = \sum_{\lambda \in \Delta_i} (\ker \pi)_\lambda$.

It is sufficient to show that these L_i are indecomposable (see Lemma 4.13).

If Δ_i is of type a, then L_i is irreducible because its weights lie in one orbit and have multiplicity 1 (see Corollary 3.14).

Obviously, there is at most one Δ_i of type b in Δ. We now use the classification of degenerate sums again, for handling the case of type b.

Note that δ^i is a short root (see 2.8 or 2.12).

<u>case 1</u>. Σ is of type F_4, p = 2.

In the irreducible representation σ of G, with heighest weight δ^i, short roots have multiplicity 1 and zero has multiplicity 2 (see [26], Table II). Comparing the multiplicities of the irreducible representation $\sigma \circ$ Fr (cf. [2], Theorem 7.5) with those of the representation in L_i, we see that L_i is irreducible again.

<u>case 2</u>. Σ is of type G_2, p = 3.

As in case 1, we see that L_i is irreducible, using ([21], 4.9).

<u>case 3</u>. Σ is of type B_1, p = 2.

In this case all multiplicities of weights in L_i are 1 (see Corollary 3.14). We have noticed earlier (see 3.10, case 1) that $Z^*_{2\varepsilon_2}$ generates a submodule having zero as a weight.

As in the proof of Lemma 4.11 it follows that L_i is indecomposable.

<u>case 4</u>. Σ is of type G_2, p = 2. Put $\alpha = \alpha_1$, $\beta = \alpha_1 + \alpha_2$, $\gamma = -\alpha-\beta$ (cf. 2.16, case 4). As p = 2, we have

$\hat{Ad}(x_{-\beta}(1))\ Z^*_{2\beta} =$

$[\hat{Ad}(x_{-\beta}(1))\ X^*_{\beta-\gamma}, \hat{Ad}(x_{-\beta}(1))\ X^*_{-\alpha}] =$

$[X^*_{\beta-\gamma} + X^*_{-\gamma} + X^*_{\alpha} + X^*_{\alpha-\beta}, X^*_{-\alpha} + X^*_{\gamma-\beta}]$.

Its component in $\mathfrak{g}^*_0$ is $H^*_{\beta-\gamma} + H^*_{\alpha}$, which is non-zero (see Corollary 2.17, (ii)).

So we can argue as in case 3, with $Z^*_{2\varepsilon_2}$ replaced by $Z^*_{2\beta}$.

Cases 1,2,3,4 cover all possiblities, whence (2). Properties (i), (ii) in the proposition follow from (1), (2). Next we prove (iv). The first irreducibility statement in (iv) has been proved above (see proof of (2)).

Now consider cases B_1, G_2; $p = 2$, δ^i short.

In L_i the weight zero has multiplicity 1, and $Z_{2\delta^i}$ generates L_i (see 3.14 and the proof of (2)).

Choose a non-zero element H^* in $(\ker \pi)_0$. (See 2.17). Calculation shows that $x_\alpha(t)$ fixes H^* for some short root α. Then H^* is fixed by all $x_\delta(t)$, δ short, because of the action of W. (W preserves $\mathfrak{g}^*_0$). If δ is a long root, then H^* is also fixed by $x_\delta(t)$, because no weight of L_i is a multiple of δ. (Use [2], Lemma 5.2).

We see that $(L_i)_0$ is a G-submodule.

The quotient $L_i/(L_i)_0$ has one orbit of weights, with multiplicity 1, hence is irreducible.

Finally, we have to prove (iii).

Let $R = L_i$, L_i as in (2).

Let $\hat{Ad}_i$ denote the restriction of $\hat{Ad}$ to L_i, and let (σ_i, L_i') denote the representation $(\rho_i)_{M_i} \circ Fr$ in L_{M_i}. (So L_i' denotes the vector space L_{M_i}, viewed as representation space of σ_i).

The highest weight δ^i of ρ_i is a minimal dominant weight (see Lemma 2.9).

So every non-zero weight of ρ_i is in the orbit of δ^i, and has multiplicity 1 (see [7], Exposé 20, Proposition 1 and Exposé 16, Proposition 4).

Suppose zero is a weight of ρ_i. Then δ^i is Σ-connected to 0 in $W\delta^i \cup (0)$, so δ^i is a multiple of a root. In fact δ^i has to be a root, because it is a minimal dominant weight.

So, if $\delta^i \notin \Sigma$ then the multiplicity of zero in ρ_i is zero. If $\delta^i \in \Sigma$, this multiplicity can be obtained from Weyl's dimension formula, or from [25]. It is seen that this multiplicity is the same as that of $\hat{Ad}_i$ in 0. (See (3.14) for the latter one).

Hence $\hat{Ad}_i$ and σ_i have the same multiplicity in zero. They also have the same multiplicity in non-zero weights.

For both representations all weight components are defined over $\mathbb{F}_p$. If $\hat{Ad}_i$ is irreducible, then it follows that $\hat{Ad}_i \cong \sigma_i$.

So we only have to consider the cases B_1, G_2 (p = 2), mentioned in (iv).

From the definition of $(\sigma_i, L_i^!)$ it follows that $L_i^!$ is generated by its heighest weight vector. Hence there is a homomorphism of G-modules $L_i^! \to L_i/(L_i)_0$, defined over $\mathbb{F}_2$.

The kernel of this homomorphism is $(L_i^!)_0$.

We see:

(3) The representations in $L_i/(L_i)_0$ and $L_i^!/(L_i^!)_0$ are isomorphic over $\mathbb{F}_2$.

(4) The representations in $(L_i)_0$ and $(L_i^!)_0$ are also isomorphic over $\mathbb{F}_2$.

One gets a T-equivariant isomorphism of vector spaces $\psi: L_i \to L_i^!$, defined over $\mathbb{F}_2$ (i.e. $h.\psi(v) = \psi(h.v)$ for $h \in T$, $v \in L_i$).

We have to show that ψ is an isomorphism of G-modules. Or, what

amounts to the same, ψ has to be an isomorphism of $\mathcal{U}_{\mathbb{Z}}$-modules. (see 4.1, formula (3)). From (3), (4) it follows that the only case that might cause any trouble is the case
$\alpha \in \Sigma$, α short, $v \in (L_i)_{-2\alpha}$, where we have to prove
$(X_\alpha^2/2) \cdot \psi v = \psi((X_\alpha^2/2) \cdot v)$.
As everything may be taken to be defined over $\mathbb{F}_2$, the problem is solved if both sides are proved to be non-zero. (Note that multiplicities are 1).
Suppose $(X_\alpha^2/2) \cdot v = 0$. Then $(X_\delta^2/2) \cdot (L_i)_{-2\delta} = 0$ for all short roots, contradicting the fact that L_i is generated by a highest weight vector. In the same way $(X_\alpha^2/2) \cdot \psi v$ is non-zero.

§6. G-invariant [p]-structures.

In this section we prove uniqueness of a [p]-structure on $\underline{g}^*$ ($\underline{g}$) that is invariant under $\hat{\text{Ad}}$ (Ad). We will see later (see Corollary 10.2) that such a [p]-structure on $\underline{g}^*$ exists. (It exists on $\underline{g}$ of course).

6.1. Recall that a [p]-structure on a Lie algebra $\underline{g}_1$ over k is a mapping $X \to X^{[p]}$ such that

(i) $\text{ad}(X^{[p]}) = (\text{ad } X)^p, \qquad (X \in \underline{g}_1)$.

(ii) $(\lambda X)^{[p]} = \lambda^p X^{[p]}, \qquad (X \in \underline{g}_1, \lambda \in k)$.

(iii) $(X+Y)^{[p]} = X^{[p]} + Y^{[p]} + \sum_1^{p-1} t_i(X,Y)$, where t_i is an expression given in [1], (3.1).

We specify t_i for p = 2,3:

$p = 2: \qquad t_1(X,Y) = [X,Y]$.

$p = 3: \qquad t_1(X,Y) = [Y,[Y,X]]$,

$\qquad\qquad t_2(X,Y) = [X,[X,Y]]$.

A Lie algebra with [p]-structure is called a p-Lie algebra.

6.2. PROPOSITION.

Assume $\Sigma \cap p\Gamma = \emptyset$.

(i) *There is at most one* [p]*-structure on* g^* *which is invariant under* $\hat{Ad}$.

(ii) *There is exactly one* [p]*-structure on* g *which is invariant under* Ad.

(iii) *If* g^* *has a* [p]*-structure as in* (i), *and* g *has the* [p]*-structure of* (ii), *then* $(\ker \pi)^{[p]} = 0$ *and* $\pi: g^* \to g$ *is a homomorphism of* p*-Lie algebras*.

PROOF.

(i) A [p]-structure is fully determined by its values on a basis. Suppose [p] is as in (i).
We shall prove that $(X_\alpha^*)^{[p]}$, $(H_\alpha^*)^{[p]}$, $(Z_\gamma^*)^{[p]}$ are computable and hence unique. If $X \in g_\beta^*$, then we have $X^{[p]} \in g_{p\beta}^*$, because of property (ii) in 6.1. It follows that $(Z_\gamma^*)^{[p]} = 0$ (γ degenerate), and $(X_\alpha^*)^{[p]} = 0$ (α a long root).
Let $\alpha, \beta, \alpha+\beta \in \Sigma$, $\alpha+\beta$ short, α long.
Then

$0 = \hat{Ad}\,(x_\beta(t))\,(X_\alpha^*)^{[p]} =$

$(X_\alpha^* + t\,\hat{ad}\,(X_\beta)\,X_\alpha^* + \sum_{j=2}^{n} t^j\, Y_j)^{[p]}$, where $Y_j \in g_{\alpha+j\beta}^*$.

So $0 = (X_\alpha^*)^{[p]} + (t\,\hat{ad}(X_\beta^*)X_\alpha)^{[p]} + \sum_{j=2}^{n} (t^j\, Y_j)^{[p]} + R$, where R is some computable expression in commutators. Taking homogeneous parts with respect to weights, we see that $-(t\,\hat{ad}\,(X_\beta)X_\alpha)^{[p]}$ is the component of R in $g_{p\alpha+p\beta}^*$.
Now

$(t\,\hat{ad}\,(X_\beta)X_\alpha^*)^{[p]} = t^p\, N_{\beta\alpha}\,(X_{\alpha+\beta}^*)^{[p]}$, and $N_{\beta\alpha} = \pm 1$.

It follows that $(X^*_{\alpha+\beta})^{[p]}$ is computable. (It is easy to check in this way, that $(X^*_{\alpha+\beta})^{[p]} = \pm Z^*_{p\alpha+p\beta}$). As every short root can be obtained in the form $\alpha+\beta$ with α long, all $(X^*_\delta)^{[p]}$ are computable ($\delta \in \Sigma$). We are done, if we prove the same for the $(H^*_\alpha)^{[p]}$. Now

$$\hat{A}d(x_{-\alpha}(t))\,(X^*_\alpha)^{[p]} = (X^*_\alpha - tH^*_\alpha + \sum_{j=2}^{n} t^j Y'_j)^{[p]} =$$

$$(X^*_\alpha)^{[p]} - t^p(H^*_\alpha)^{[p]} + \sum_{j=2}^{n} (t^j\; Y'_j)^{[p]} + R', \text{ where } Y'_j \in \underline{g}^*_{\alpha-j\alpha},$$

R' is computable.

Taking homogeneous parts again, we see that $(H^*_\alpha)^{[p]}$ is computable.

(ii) The uniqueness is proved in the same way as for $\underline{g}^*$, the existence follows from the fact that $\underline{g}$ is the Lie algebra of the algebraic group G (see [1], (3.3)).

(iii) We have proved $(Z^*_\gamma)^{[p]} = 0$, γ degenerate sum. So we still have to prove that $((\ker\pi)_0)^{[p]} = 0$. As $\ker\pi$ is abelian we have for short roots α:

$$0 = \hat{A}d\,(x_{-\alpha}(t))\,(Z^*_{p\alpha})^{[p]} = (Z^*_{p\alpha} + t^p Z_0 + t^{2p} Z_1)^{[p]} =$$

$$0 + t^{p^2}\, Z_0^{[p]} + 0, \text{ where } Z_i \in (\ker\pi)_{-ip\alpha}.$$

The elements of type Z_0 span $(\ker\pi)_0$ (see Proposition 5.2). It follows that $(\ker\pi)^{[p]} = 0$.

Now we define a [p]-structure on $\underline{g}$ by the relation:

$$(\pi\, X)^{[p]} = \pi\, X^{[p]}.$$

If $\pi X = \pi Y$, then $X-Y$ is central, $(X-Y)^{[p]} = 0$, so $X^{[p]} = (Y+(X-Y))^{[p]} = Y^{[p]}$. Hence [p] is well-defined on $\pi(\underline{g}^*) = \underline{g}$. It is invariant under Ad, so it is the [p]-structure of (ii).

6.3. REMARKS.

1) From the proof of (i) it follows that the [p]-structure of (iii)

is defined over $\mathbb{F}_p$.

2) Suppose that $\underline{g}^*$ has a [p]-structure as in (i). Then $(H^*_\alpha)^{[p]} = H^*_\alpha$ for long roots α, because the computation of $(H^*_\alpha)^{[p]}$ is "the same" as the computation of $(H_\alpha)^{[p]}$.

But $(H^*_\alpha)^{[p]} \neq H^*_\alpha$ for α short. For suppose $(H^*_\alpha)^{[p]} = H^*_\alpha$ is also true for short roots.

Then $(H^*)^{[p]} = H^*$ for all $H^* \in (\underline{g}^*)_0$, hence for all $H^* \in (\ker \pi)_0$. This contradicts (ii). (see Corollary 2.17). If α is a short root then H^*_α is not a semisimple element but the sum of a semisimple part, spanned by the H^*_β with β long, and a nilpotent part in $\ker \pi$. (See [20], p. 119 for definitions).

3) For α long $(X^*_\alpha)^{[p]} = 0$, but for α short $(X^*_\alpha)^{[p]} \neq 0$. One reason for this inequality is that otherwise the computation of $(H^*_\alpha)^{[p]}$ would not differ from the computation of $(H_\alpha)^{[p]}$, which would contradict remark 2.

4) The existence of a [p]-structure as in (i) can be proved along the same line as the uniqueness. We don't need this method. (See section 10).

§7. The extension $\phi : G^* \to G$.

We look for an interpretation of $\pi : \underline{g}^* \to \underline{g}$ as the differential $d\phi$ of a homomorphism ϕ of algebraic groups (see [1], (3.3)). In this section we make some remarks about such a homomorphism.

We suppose that the codomain of ϕ is an almost simple Chevalley group G, having $\underline{g}$ as its Lie algebra.

Let G^* denote the domain of ϕ. If ϕ is such that $d\phi$ is a universal central extension of $\underline{g}$, then the restriction of ϕ to the connected component of G^* also has that property. Hence we suppose that G^* is connected. In 2.1 we only considered the case that G is simply connected. We give a justification for that choice now.

7.1. LEMMA.

<u>Let</u> G <u>be an almost simple Chevalley group, with Lie algebra</u> $\underline{g}$, <u>such that</u>

(i) $\underline{g} = [\underline{g}, \underline{g}]$,

(ii) $\underline{g} \neq \underline{g}^*$.

<u>Then</u> G <u>is simply connected</u>.

PROOF. Let G_1 be the simply connected Chevalley group that covers G, and let $\underline{g}_1$ be its Lie algebra. We claim that the natural homomorphism τ: $\underline{g}_1 \to \underline{g}$ is an isomorphism.
It is well known that the image of τ contains all $\underline{g}_\alpha$, $\alpha \in \Sigma$ (see [2], 2.6).
From (i) it follows that $\underline{g}$ is generated by these $\underline{g}_\alpha$. So τ is surjective. Then τ is an isomorphism, because dim $\underline{g}$ = dim $\underline{g}_1$. We may conclude that $\underline{g}_1 = [\underline{g}_1, \underline{g}_1]$, $\underline{g}_1^* \neq \underline{g}_1$.
This situation was analysed before. We see that there are degenerate sums and that the order of Γ/Γ_0 is a power of p (see Corollary 3.14 and Lemma 2.10). The Lie algebras $\underline{g}$, $\underline{g}_1$ are obtained from lattices M, M_1 in $\underline{g}_{\mathbb{C}}$, with $M \supset M_1$. The group M/M_1 is isomorphic to a subgroup of Γ/Γ_0. (see [2], 2.6).
So its order is a power of p. But $L_{M/M_1} = 0$ because τ is surjective.
It follows that $M = M_1$, hence $G \cong G_1$.

REMARK. If we don't require that G is almost simple, then the proof shows that $\underline{g}$ is the direct sum of the Lie algebras of the almost simple factors of G. Then it is easy to see that $\dot{\pi} : \underline{g}^* \to \underline{g}$ is the direct sum of the corresponding universal central extensions.

7.2. We return to the notations p, $\underline{g}$, G, .. of 2.1. Suppose that there is a homomorphism ϕ: $G^* \to G$ as above, that is, such that G^* is a connected algebraic group and $d\phi$ is a universal central extension of $\underline{g}$. The Lie algebra of G^* can be identified with $\underline{g}^*$.

Then $d\phi$ is identified with π.

We will henceforth indicate this situation by the statement

(1) $d\phi = \pi$.

Assume (1).

The Lie algebra $\underline{g}^*$ has a [p]-structure, which is invariant under $Ad: G^* \to Aut(\underline{g}^*)$.

As π is surjective, ϕ is also surjective.

For $x \in G$, choose $y \in G^*$, such that $\phi y = x$.

Then $\pi \circ Ad(y) = Ad(x) \circ \pi$, hence $Ad(y) = \hat{Ad}(x)$.

We see that the [p]-structure on $\underline{g}^*$ is invariant under $\hat{Ad}$.

So it is the [p]-structure discussed in 6.2.

7.3. The Lie subalgebra $\ker \pi$ of $\underline{g}^*$ is an abelian Lie algebra with trivial [p]-structure (see Proposition 6.2). So $(\ker \phi)^0$, i.e. the connected component of $\ker \phi$, is the unipotent radical R_u of G (see [1], Cor. (8.2), (11.5)). In fact we have:

7.4. LEMMA.

$$R_u = \ker \phi$$

PROOF.

G^*/R_u is connected, and there is a separable homomorphism $\psi: G^*/R_u \to G$. The group G is simply connected. The group G^*/R_u has the same dimension as G, the same semisimple rank, the same root system. We see that there is an inverse for ψ, or that ψ is an isomorphism. (See [7], Exposé 23, Théorème 1). So

$\ker \phi = \ker (G^* \to G^*/R_u \to G) = \ker (G^* \to G^*/R_u) = R_u$.

7.5. Now let G^* be a connected algebraic group with Lie algebra $\underline{g}^*$, $\underline{g}^* \neq \underline{g}$. Suppose that the [p]-structure of $\underline{g}^*$ is invariant under $\hat{Ad}$. Let R_u be the unipotent radical of G^*, with Lie algebra $\underline{r}_u$.

Then $\underline{g}' = \underline{g}^*/\underline{r}_u$ is the Lie algebra of the reductive group $G' = G^*/R_u$.

This group G' is its own commutator, because $\underline{g}' = [\underline{g}', \underline{g}']$ (see [1], (3.12)). So G' is even a semi-simple group. Now we use the following lemma.

LEMMA.

In the Lie algebra $\underline{g}'$ of a reductive algebraic group G' there is no central nilpotent element.

PROOF. Let $\underline{c}$ be the set of central nilpotent elements. It is an ideal, invariant under Ad. It has no weight space with weight zero, because $\underline{g}_0'$ consists of simi-simple elements. Let $\underline{c}_\alpha$ be a weight space of $\underline{c}$. Then α is a root, so $\underline{c}_\alpha = \underline{g}_\alpha'$ and $\underline{c}_\alpha$ is contained in the Lie algebra of a subgroup of type SL_2 or PSL_2. Hence it is sufficient to prove the Lemma for SL_2 and PSL_2, which is easy.

7.6. Applying the Lemma, we see that the image of ker π in $\underline{g}'$, which consists of central nilpotent elements, is zero. So

(1) $\ker \pi \subset \underline{r}_u$.

Let $\underline{i}$ denote the image $\pi(\underline{r}_u)$ of $\underline{r}_u$ in $\underline{g}$. It is an ideal that consists of nilpotent elements.

We have $\underline{i} = \sum_{(\alpha)} \underline{i}_{(\alpha)}$, where (α) runs over local weights.

(i.e. $\underline{i}_{(\alpha)} = \{X \in \underline{i} \mid [H_\beta, X] = \langle\alpha,\beta\rangle X$, for all $\beta \in \Sigma\}$. Local weights are elements of $\Gamma/p\Gamma$, global weights are elements of Γ).

The term $\underline{i}_{(0)}$ is zero, because $\underline{h} = \underline{g}_{(0)}$ consists of semi-simple elements. (Recall that $\Sigma \cap p\Gamma = \emptyset$). So if β is a global root, that behaves like $(-\alpha)$ locally, then $[X_\beta, \underline{i}_{(\alpha)}] = 0$. On the other hand $[X_\beta, \underline{g}_{-\beta}] \neq 0$. We see that $\underline{i} = 0$. Together with (1) this proves

(2) $\underline{r}_u = \ker \pi$.

So $\underline{g}' \cong \underline{g}^*/\underline{r}_u \cong \underline{g}$.

Hence $\underline{g}'$ is not the direct sum of two proper subalgebras, and G' is almost simple. (Use the remark in 7.1).
Then G' is isomorphic to a Chevalley group over K (see [7], cf. [2], 3.3(6)), so we can apply Lemma 7.1. We see that G' is isomorphic to a simply connected almost simple Chevalley group with the same rank and the same dimension as G.
Then it follows that $G \cong G'$ (use 2.8, Table 1).

7.7. We conclude from the above: Over K the following two problems are equivalent:
(i) To find a homomorphism ϕ such that $d\phi = \pi$.
(ii) To find an algebraic group G^* which has $\underline{g}^*$ as its Lie algebra, such that the [p]-structure on $\underline{g}^*$ is invariant under $\widehat{Ad}$.

REMARK.
We shall use the first formulation in our solution.

7.8. DEFINITION.
Let G,H be connected linear algebraic groups, $\phi: H \to G$ a homomorphism of algebraic groups such that $d\phi$ is a central extension (so ϕ is surjective and separable). Then ϕ is called an <u>infinitesimally central extension</u> of G.

7.9. Consider an infinitesimally central extension $\psi: H \to G$ where G is a Chevalley group and H is a linear algebraic group with perfect Lie algebra $\underline{h}$ (i.e. $\underline{h} = [\underline{h}, \underline{h}]$). It follows from Proposition 1.3, (v) that there is a surjective homomorphism of Lie algebras $\rho: \underline{g}^* \to \underline{h}$. It is easy to see that ρ is a universal central extension. Analogously to the problem of finding a homomorphism ϕ with $d\phi = \pi$ (as in 7.2), there is the problem of finding a homomorphism χ such that $d\chi = \rho$. This last problem will be discussed in section 13 (see Theorem 13.9). Note that such a homomorphism χ is an infinitesimally central extension and that the same is true for $\psi \circ \chi$.

§8. Extensions of G by a G-module.

In this section we discuss extensions of a group G by a G-module V.

8.1. Let G be a connected algebraic group, defined over k. Let V be a (finite dimensional) G-module over k (i.e. V is defined over k and the action is defined over k).

NOTATIONS.

The semi-direct product of G and V is denoted $\ulcorner V,G \urcorner$, and its elements are denoted $\ulcorner v,g \urcorner$.

So $\ulcorner v,g \urcorner \ulcorner v',g' \urcorner = \ulcorner v+g\cdot v', gg' \urcorner$.

The projections $\ulcorner V,G \urcorner \to V$, $\ulcorner V,G \urcorner \to G$, and the injections $V \to \ulcorner V,G \urcorner$, $G \to \ulcorner V,G \urcorner$ are denoted p_V, p_G, i_V, i_G respectively.

Let V' be another G-module, and ϕ: $V \to V'$ a homomorphism of G-modules. Let $\psi : G \to G$, $\chi : G \to V$ be morphisms. Then we denote $\ulcorner \phi,\psi \urcorner$ the morphism that sends $\ulcorner v,g \urcorner$ to $\ulcorner \phi v,\psi g \urcorner$, and $\ulcorner \chi,\psi \urcorner$ the morphism that sends g to $\ulcorner \chi g,\psi g \urcorner$.

DEFINITION.

Let G act on two varieties X and Y, and let f: $X \to Y$ be a morphism. Then f is called <u>G-equivariant</u> if $g\cdot f(x) = f(g\cdot x)$ for all $g \in G$, $x \in X$.

DEFINITION.

An <u>extension</u> of G by the G-module V is a homomorphism ϕ: $H \to G$ with the following properties

(i) ϕ is surjective and $d\phi$ is surjective. (So ϕ is separable and $G \cong H/\ker \phi$. See [1], (6.6)).

(ii) ker ϕ is abelian. (So the represention Int of H in ker ϕ factors through G).

(iii) There is a G-equivariant isomorphism of algebraic groups $\tau : V \to \ker \phi$.

We say that $\phi: H \to G$ is a k-extension if H is defined over k (i.e. H is a k-group) and ϕ,τ are defined over k.

8.2. THEOREM. (Existence of a T-equivariant cross-section). Let $\phi: H \to G$ be a k-extension of G by V,T a k-split maximal torus of H. Then there is a morphism $s : G \to H$, defined over k, such that $\phi \circ s = \mathrm{id}$ and

(i) $s(\phi T) = T$

(ii) $\mathrm{Int}(t)(s(g)) = s(\mathrm{Int}(\phi t)(g))$ for $t \in T$, $g \in G$.

(So s is T-equivariant).

PROOF.

In fact we will only need the structure of V as a T-module, not the structure of V as a G-module. First we use the method of [3], 9.5, to get a T-equivariant cross-section s, defined over k. There has to be made a slight modification in the proof of loc. cit. One has to put $s': x \mapsto c(x) \cdot s(x)$ instead of $s': x \mapsto s(x) \cdot c(x)$. With this modification the proof also works in our case. We get a cross-section s that satisfies (ii).

We have to change s in such a way that it also satisfies (i). Hence we look for a T-equivariant morphism $r: G \to V$, defined over k, such that

(1) $r(\phi(t))s(\phi(t)) = t$ for all $t \in T$.

If r exists, then rs satisfies both (i) and (ii) and we are done. The restriction of ϕ to T is an isomorphism to ϕT, because ϕ is separable and ker ϕ is unipotent. Let ψ be the inverse of this isomorphism. Then (1) can be written as:

(2) $r(t) = \psi(t)s(t)^{-1}$ for all $t \in \phi T$.

The righthand side of (2) is a morphism $r' : \phi T \to V$, defined over k, that is T-equivariant, hence it maps ϕT into the weight space V_0. We claim that it can be extended to a T-equivariant morphism $r : G \to V_0$, defined over k.

For such a morphism r the T-equivariance means

(3) $r \circ \mathrm{Int}(t) = r$ for all $t \in \phi T$.

So consider the representation of the k-split torus ϕT in the affine algebra $A[G]$ of G, defined by $t.f = f \circ \mathrm{Int}(t)$. (So $(t.f)(x) = f(\mathrm{Int}(t)(x))$ for $x \in G$).

This representation is defined over k, and each f is contained in a finite-dimensional subspace, stable under ϕT. Hence we have a decomposition into weight spaces. If $A[\phi T]$ is the affine algebra of ϕT, then $\tau : A[G] \to A[\phi T]$ is surjective, defined over k. Let ϕT act trivially on $A[\phi T]$.

Then τ is also a homomorphism of ϕT-modules. ($\tau(f)$ is the restriction of f to ϕT). We conclude that $\tau(A_k[G]_0) = A_k[\phi T]$.

It follows that the righthand side r' of (2) can be extended to a morphism r, defined over k, satisfying (3).

REMARK

The condition "T is k-split" can be dropped.

We only need that T is defined over k, because it can be proved without the assumption about the splitting that the weight spaces V_0, $A[G]_0$ are defined over k (see [1], 9.2, Corollary).

§9. The Hochschild groups.

We will use rational cohomology to describe $\phi : G^* \to G$ (see 7) as an extension of G. In this section we recall some facts about this cohomology (cf. [11], Ch. II, §3).

9.1 DEFINITIONS AND NOTATIONS.

Let G be a connected algebraic group, defined over k. Let V be a (finite dimensional) G-module over k.

A (regular) n-cochain of G in V is a morphism $G \times \ldots \times G \to V$, where $G \times \ldots \times G$ denotes the direct product of n copies of the variety G. (If n = 0, then this product consists of 1 point).

We put

$$C^n(G,V) = \{\text{n-cochains of G in V}\},$$

$$C^n_k(G,V) = \{\text{n-cochains of G in V, defined over k}\}.$$

The set $C^n(G,V)$ can be viewed as a vector space in a natural way. The subset $C^n_k(G,V)$ is a k-structure on this vector space.

The boundary operator $\partial^n : C^n(G,V) \to C^{n+1}(G,V)$ is defined by

$$(\partial^n f)(g_1,\ldots,g_{n+1}) = g_1.f(g_2,\ldots,g_{n+1}) +$$
$$\sum_{i=1}^{n} (-1)^i f(g_1,\ldots,g_i g_{i+1},\ldots,g_{n+1}) + (-1)^{n+1} f(g_1,\ldots,g_n).$$

The boundary operator is defined over k.

The n-th Hochschild group of G in V is the group $H^n(G,V) = (\ker \partial^n)/(\operatorname{Im} \partial^{n-1})$. It is denoted $H^n(V)$ if no confusion is possible. An element of $\ker \partial^n$ is called an n-cocycle, an element of $\operatorname{Im} \partial^{n-1}$ is called an n-coboundery.

The class mod$(\operatorname{Im} \partial^{n-1})$ of an n-cocycle f is denoted $\overline{f}$. Let ∂^n_k denote the restriction of ∂^n to $C^n_k(G,V)$. Then we put

$$H^n_k(G,V) = (\ker \partial^n_k)/(\operatorname{Im} \partial^{n-1}_k).$$

It is also denoted $H^n_k(V)$, and it may be identified with the k-structure of $H^n(V)$, consisting of classes mod$(\operatorname{Im} \partial^{n-1})$ that meet $C^n_k(G,V)$. It is easy to see that:

9.2 LEMMA.

If k' *is a field extension of* k, *then*

$$H^n_{k'}(V) \cong H^n_k(V) \otimes_k k'.$$

9.3 We want to give interpretations for n-cocycles and n-coboundaries, $n \leq 2$ (cf. [11], Ch. II, §3 or [5], Ch. X §4, Ch. XIV §4).

<u>n = 0</u> The only coboundary is 0. The cocycles correspond to elements v of V, fixed by G. They are called invariants.

<u>n = 1</u> Let $\ulcorner V,G \urcorner$ be the semi-direct product of V and G (see 8.1). Every 1-cochain f defines a section $s : x \mapsto \ulcorner f(x),x \urcorner$ of p_G. This section is a homomorphism if and only if f is a cocycle.

<u>n = 2</u> Let $\phi : H \to G$ be a k-extension of G by V. Then there is a section s of ϕ, defined over k. (See [19], Corollary 1 to Theorem 1, or the remark in 8.2). So H is isomorphic to the variety $V \times G$ by means of $x \mapsto (x(s\phi x)^{-1}, \phi x)$. We transfer the group structure to $V \times G$ by means of this isomorphism.
Put $f(x,y) = s(x)s(y)s(xy)^{-1}$.
Then $(v,g)(v',g') = (v+g.v' + f(g,g'),gg')$ in $V \times G$ and f is a 2-cocycle. Every 2-cocycle can be obtained in this way. Two 2-cocycles differ a coboundary if and only if they are obtained from isomorphic extensions. (Or from two sections in the same extension).

9.4 Let $\mathcal{E} : 0 \to A \xrightarrow{\tau} B \xrightarrow{\rho} C \to 0$ be an exact sequence of G-modules over k. Then there is a long exact sequence

$$0 \to H^0(A) \xrightarrow{H^0(\tau)} H^0(B) \xrightarrow{H^0(\rho)} H^0(C) \xrightarrow{\delta^0(\mathcal{E})} H^1(A) \xrightarrow{H^1(\tau)}$$
$$H^1(B) \xrightarrow{H^1(\rho)} H^1(C) \xrightarrow{\delta^1(\mathcal{E})} H^2(A) \;\ldots\ldots$$

where the connecting homomorphisms $\delta^i(\mathcal{E})$, also denoted δ^i, may be defined as follows:
Choose a section σ of ρ, compatible with the linear structures. (In fact we only need that σ is a morphism of varieties such that $\sigma \circ \rho = \mathrm{id}$. We just make a better choice here.)

Let f be an i-cocycle in C. Then $\sigma \circ f$ is an i-cochain in B. The (i+1)-coboundary $\partial^i(\sigma \circ f)$ has its values in A. So it is an (i+1)-cocycle in A.(It is not necessarily a coboundary in A). Its class in $H^{i+1}(A)$ is $\delta^i(\overline{f})$.

9.5 EXAMPLE.

Let i = 1 and let f be a 1-cocycle in C. To f corresponds a section $s : G \to \ulcorner C,G \urcorner$ of p_G. ($s = \ulcorner f,id \urcorner$).
Let ψ be the natural homomorphism $\ulcorner \rho,id \urcorner : \ulcorner B,G \urcorner \to \ulcorner C,G \urcorner$. Then an element of $\delta^1(\overline{f})$ corresponds to an extension that is isomorphic to the extension $p_G \circ \psi : \psi^{-1}(sG) \to G$. (Note that $p_G \circ \psi$ is the p_G of $\ulcorner B,G \urcorner$). That extension is a subextension, with kernel A, of $\ulcorner B,G \urcorner \to G$.
One may take as section of $p_G \circ \psi$ the morphism $\ulcorner \sigma \circ f,id \urcorner$. ($\sigma : C \to B$ as above).

9.6 Now we return to the case that G is a simply connected Chevalley group, defined over k, where k is a field of characteristic $p > 0$.

THEOREM. (*cf. Steinberg* [23]).
Let L *be a* G-*module over* k, *on which* G *acts trivially. Then* $H^2_k(G,L) = 0$.

PROOF.

We may assume that k is the algebraic closure of $\mathbb{F}_p$, because of Lemma 9.2.Let f be a 2-cocycle, defined over k. There corresponds to f a k-extension $\phi : H \to G$ of G by L, with section s.
Now some well-known results of Steinberg (see [23], Th. 3.2, 3.3, 4.1) show that there is a homomorphism $\psi : G(k) \to H(k)$

with $\phi \circ \psi = \mathrm{id}$. We shall show that ψ is a morphism. Then ψ is a section of ϕ that defines the cocycle 0, hence $\overline{f} = 0$.

As ϕ is a central extension (i.e. ker ϕ is in the centre of H), we have

(1) $\psi((x,y)) = (\psi(x),\psi(y)) = (s(x),s(y))$ for $x,y \in G(k)$.

(Central trick for groups, cf. 1.2. See 2.1 for notations). Now G(k) is its own commutator group. (As k is algebraically closed, this follows from $\underline{g} = [\underline{g},\underline{g}]$. It is true in the general case too. See [2], 3.3 (5)).

So (1) determines ψ.

Take $a \in k^{\times}$ such that $a^2 \neq 1$. (cf. [23], 9.1). Let $\alpha \in \Sigma$, $t \in k$. Then $\psi(x_\alpha(t)) = \psi((h_\alpha(a), x_\alpha((a^2-1)^{-1}t))) = (sh_\alpha(a), sx_\alpha((a^2-1)^{-1}t))$. (See 2.1 for notations). We see that the restriction of ψ to $\{x_\alpha(t) | t \in k\}$ is a morphism. It follows that the restriction to $\{h_\beta(t) | t \in k^\times\}$ is also a morphism. ($\beta \in \Sigma$). Then the restriction to the open cell (see (2.1)) is a morphism, because the open cell is the direct product (as a variety) of the subgroups $\{x_\alpha(t) | t \in k\}$, $\alpha \in \Sigma$, and $\{h_\beta(t) | t \in k^\times\}$, β simple (see [2], 3.3 (3) and [8], Proposition 1). By right translation we see that ψ is a morphism locally, hence ψ is a morphism.

§10. The existence of $\phi : G^* \to G$.

We now return to the problem of finding $\phi : G^* \to G$ such that $d\phi = \pi$ (see 7.2). In this section we give a constructive proof of the existence of ϕ. Uniqueness will be discussed later, in section 13.

Let $G, \underline{g}^*, \pi, p, T, \hat{A}d, \ldots$ be as in 2.1, 3.1, 3.4.

NOTATION.

The G-module ker π, that is described in 5.2 is denoted $\underline{r}_u$.

10.1 THEOREM.

Assume $\Sigma \cap p\Gamma = \emptyset$.

There is a k-extension $\phi : G^* \to G$ of G by $\underline{r}_u$, such that $d\phi$ is a universal central extension of $\underline{g}$.

10.2 COROLLARY.

If $\Sigma \cap p\Gamma = \emptyset$, then there is exactly one [p]-structure on $\underline{g}^*$ that is invariant under $\hat{A}d$ (see 6.2 and 7.2).

10.3 PROOF OF THE THEOREM. (This proof is lengthy).

We may assume that $\underline{r}_u \neq 0$, or, equivalently, that degenerate sums exist. Constructions of ϕ will be given type by type, using the classification of degenerate sums.

First we describe the general method that underlies these constructions. To get the extension of G by $\underline{r}_u$, we look for a suitable 2-cocycle f_2 of G in $\underline{r}_u$. We now describe how this 2-cocycle is obtained and how it is checked whether it is suitable.

$\underline{1^0}$ (SKETCHY)

Let

$\mathcal{E}_1 : 0 \to \underline{r}_u \xrightarrow{\mu} C \xrightarrow{\nu} A \to 0$ and

$\mathcal{E}_2 : 0 \to L_1 \xrightarrow{\rho} A \xrightarrow{\sigma} B \xrightarrow{\tau} L_2 \to 0$ be exact sequences of G-modules over k, such that G acts trivially on L_1, L_2, $\dim L_2 = 1$. Take a non-zero element of $(L_2)_k$. It corresponds to a 0-cocycle f_0 of G in L_2, defined over k. Using the short exact sequence $\mathcal{E}_{2,2} : 0 \to \ker\tau \to B \to L_2 \to 0$, we get an element

$\delta^0(\mathcal{E}_{2,2})(\overline{f}_0)$ of $H^1_k(\ker\tau)$.

The sequence

$0 \to L_1 \to A \to \ker\tau \to 0$ is exact, so the sequence

$H^1(L_1) \to H^1(A) \to H^1(\ker\tau) \to H^2(L_1)$ is exact.

As $H^2(L_1) = 0$ (see Theorem 9.6), there is an element $\overline{f}_1$ of $H^1(A)$ that is mapped to $\delta^0(\mathcal{E}_{2,2})(\overline{f}_0)$. In fact $\overline{f}_1$ is unique, because $H^1(L_1) = 0$ too. (This follows from the fact that G is its own commutator subgroup.)
Now we choose $f_2 \in \delta^1(\mathcal{E}_1)(\overline{f}_1)$, and check whether f_2 is suitable, i.e. whether f_2 defines an extension ϕ such that $d\phi$ is a universal central extension.

<u>2^0</u> (ELABORATE).

There is some freedom in the choice of representatives and in the way they are constructed. In order to be able to check whether f_2 is suitable, we will make these choices in a convenient way.
We start with f_0 again,
(1) $f_0 \in C^0_k(G,L_2)$, corresponding to an element of $(L_2)_k$ that we also denote f_0. Choose a T-equivariant linear section η_1 of τ, defined over k. (τ occurs in the sequence $\mathcal{E}_2$). So
(2) $\tau \circ \eta_1 = \mathrm{id}$. Put
(3) $l_1 = \partial^0(\eta_1 f_0)$. It is a representative of $\delta^0(\mathcal{E}_{2,2})(\overline{f}_0)$ in $H^1_k(\ker\tau)$. (See 9.4). So it corresponds to a homomorphism $\ulcorner l_1,\mathrm{id}\urcorner : G \to \ulcorner\ker\tau, G\urcorner$, defined over k. Let $x \in G$, $h \in T$. Then
$\ulcorner l_1,\mathrm{id}\urcorner(hxh^{-1}) = \ulcorner(hxh^{-1}).\eta_1 f_0 - \eta_1 f_0,\ hxh^{-1}\urcorner =$
$\ulcorner(hx).\eta_1(h^{-1}.f_0) - \eta_1(h.f_0),\ hxh^{-1}\urcorner =$
$\ulcorner h.(x.\eta_1 f_0) - h.(\eta_1 f_0),\ hxh^{-1}\urcorner$.
We see that $\ulcorner l_1,\mathrm{id}\urcorner$ is T-equivariant, if T acts on G by Int and on $\ulcorner\ker\tau,G\urcorner$ by $\mathrm{Int}\circ i_G$. (i_G is defined in 8.1).
Now we look for a 1-cocycle f_1 in A, such that
(4) $\sigma \circ f_1 = l_1$. (Recall that $\sigma : A \to B$.)
Equivalently, we look for a homomorphism $\ulcorner f_1,\mathrm{id}\urcorner : G \to \ulcorner A,G\urcorner$ such that $\ulcorner\sigma,\mathrm{id}\urcorner\ulcorner f_1,\mathrm{id}\urcorner = \ulcorner l_1,\mathrm{id}\urcorner$.

Choose a T-equivariant linear section $\eta_2 : \ker \tau \to A$ of σ, defined over k. So

(5) $\sigma \circ \eta_2 = \mathrm{id}$.

Let G_1 denote the image of $\ulcorner l_1, \mathrm{id} \urcorner$ and let H_1 denote its inverse image in $\ulcorner A,G \urcorner$. Then $\ulcorner \sigma, \mathrm{id} \urcorner : H_1 \to G_1$ is a central extension with kernel $\ulcorner L_1, 1 \urcorner$. As $\ulcorner l_1, \mathrm{id} \urcorner$ is an isomorphism, defined over k, the group G_1 is k-isomorphic to G. Furthermore H_1 is the image of the morphism $L_1 \times G \to H_1$, defined by $(v,g) \mapsto \ulcorner v,1 \urcorner \ulcorner \eta_2 \circ l_1(g), g \urcorner$. This morphism is a k-isomorphism. So H_1 is also defined over k. We see that $\ulcorner \sigma, \mathrm{id} \urcorner : H_1 \to G_1$ is a k-extension. Then it follows from Theorem 9.6 that $H_1 \to G_1$ is isomorphic over k to the trivial extension $L_1 \times G_1 \to G_1$, where $L_1 \times G_1$ denotes the direct product of groups. So there is a homomorphism $\psi_1 : G_1 \to H_1$, defined over k, such that

(6) $\ulcorner \sigma, \mathrm{id} \urcorner \circ \psi_1 = \mathrm{id}$.

Now we choose f_1 such that

(7) $\ulcorner f_1, \mathrm{id} \urcorner = \psi_1 \circ \ulcorner l_1, \mathrm{id} \urcorner$.

Then f_1 is defined over k, and it follows from $\ulcorner \sigma, \mathrm{id} \urcorner \circ \ulcorner f_1, \mathrm{id} \urcorner =$ $= \ulcorner l_1, \mathrm{id} \urcorner$ that f_1 satisfies (4). We claim that

(8) f_1 is T-equivariant.

(9) $f_1(x_\alpha(t)) = \eta_2 \circ l_1(x_\alpha(t))$ for $\alpha \in \Sigma$, $t \in k$.

(10) $f_1(T) = 0$.

Proofs:

For $h \in T$ we have $l_1(h) = h.\eta_1 f_0 - \eta_1 f_0 = 0$. So the image T_1 of T in G_1 is $i_G T = \ulcorner 0,T \urcorner$. (Recall that $G_1 = \ulcorner l_1, \mathrm{id} \urcorner G$). Let 0_A denote the zero element in A, 0_B the zero element in B. Then $\ulcorner L_1, 1 \urcorner$ is unipotent and commutes with $\ulcorner 0_A, T \urcorner$. So $\ulcorner 0_A, h \urcorner$ is the semisimple part of $\psi_1 \ulcorner 0_B, h \urcorner$ for $h \in T$.

It follows that $\ulcorner f_1,id\urcorner(h) = \psi_1\ulcorner 0_B,h\urcorner = \ulcorner 0_A,h\urcorner$. This proves (10).
It also follows from $\psi_1\ulcorner 0_B,h\urcorner = \ulcorner 0_A,h\urcorner$ that ψ_1 is T-equivariant.
We have seen above that $\ulcorner l_1,id\urcorner$ is T-equivariant, hence $\ulcorner f_1,id\urcorner$ is T-equivariant. That proofs (8).
Now let $f_1(x_\alpha(t)) = \sum_{i=1}^{n} t^i v_i$, $v_i \in A$. (f_1 is a morphism with $f_1(1) = 0$). We have for $h \in T$:

$$\sum_{i=1}^{n} t^i(h.v) = h.f_1(x_\alpha(t)) = f_1(x_\alpha(h^\alpha t)) = \sum_{i=1}^{n} t^i h^{i\alpha} v_i.$$

(h^γ denotes the image of h under γ). It follows that $v_i \in A_{i\alpha}$.
The kernel of σ is contained in the weight space A_0, so the restriction of σ to $\bigoplus_{i>0} A_{i\alpha}$ is an isomorphism, with inverse η_2.
Hence $f_1(x_\alpha(t)) = \eta_2 \circ \sigma \circ f_1(x_\alpha(t)) = \eta_2 \circ l_1(x_\alpha(t))$.
That proves (9).

Finally, we choose a T-equivariant linear section η_3 of ν, defined over k. So

(11) $\nu \circ \eta_3 = id$.

We put

(12) $f_2 = \partial^1(\eta_3 \circ f_1)$.

(13) Then f_2 is a 2-cocycle in $\underline{r}_u$, defined over k, corresponding to the k-extension $\phi : G^* \to G$, where G^* is the inverse image of $\ulcorner f_1,id\urcorner G$ under the map $\ulcorner \nu,id\urcorner : \ulcorner C,G\urcorner \to \ulcorner A,G\urcorner$, and ϕ is the restriction to G of $p_G : \ulcorner C,G\urcorner \to G$. (See Example 9.5). It is seen as above (see proof of (6)) that ϕ is a k-extension.

Put

(14) $s = \ulcorner \eta_3 \circ f_1,id\urcorner$.

Then s is a morphism as in Theorem 8.2. We have to check whether $d\phi$ is a universal central extension. Of course, the result depends on $\mathcal{E}_1,\mathcal{E}_2$. First we prove that $d\phi$ is a central extension.

Put

(15) $R_u = \ulcorner \underline{r}_u,1\urcorner$.

Then R_u is the unipotent radical of G^*. We identify its Lie algebra with $\underline{r}_u$. The action of G on R_u is of the form $\rho \circ Fr$, where ρ is a rational representation (see Proposition 5.2). As $d(Fr) = 0$, this action of G on R_u induces a trivial action of $\underline{g}$ on $\underline{r}_u$.

In formula: $d(Ad \circ s)(\underline{g})(\underline{r}_u) = 0$.

The Lie algebra $\underline{g}_1^*$ of G^* is the direct sum, as a vector space, of $\underline{r}_u$ and $(ds)\underline{g}$, because $(v,g) \mapsto vs(g)$ is an isomorphism of varieties $R_u \times G \to G^*$. So $ad(\underline{g}^*)(\underline{r}_u) = ad((ds)\underline{g})(\underline{r}_u)+ad(\underline{r}_u)(\underline{r}_u) = d(Ad \circ s)(\underline{g})(\underline{r}_u) = 0$. This proves that $d\phi : \underline{g}_1^* \to \underline{g}$ is a central extension. (Its kernel is $\underline{r}_u$.)

So we have a homomorphism $\underline{g}^* \to \underline{g}_1^*$ with image $[\underline{g}_1^*,\underline{g}_1^*]$. (See Proposition 1.3, (v)). Now suppose $\underline{g}_1^* = [\underline{g}_1^*,\underline{g}_1^*]$. Then $\underline{g}^* \to \underline{g}_1^*$ is a surjective isomorphism, because dimensions are equal. We conclude:

(16) If $[\underline{g}_1^*,\underline{g}_1^*] = \underline{g}_1^*$, then $d\phi$ is a universal central extension.

Note that this condition is also necessary.

One has $d\phi[\underline{g}_1^*,\underline{g}_1^*] = [\underline{g},\underline{g}] = \underline{g}$. So $[\underline{g}_1^*,\underline{g}_1^*] = \underline{g}_1^*$ if and only if $\underline{r}_u$ is contained in $[\underline{g}_1^*,\underline{g}_1^*]$. Hence we consider $\underline{r}_u \cap [\underline{g}_1^*,\underline{g}_1^*]$. It is a G-submodule of $\underline{r}_u$, because both $\underline{r}_u$ and $[\underline{g}_1^*,\underline{g}_1^*]$ are invariant under $Ad \circ s$. Consider the following condition:

(17) $[(ds)\underline{g},(ds)\underline{g}] \cap \underline{r}_u$ generates $\underline{r}_u$ as a G-module.

As $[(ds)\underline{g},(ds)\underline{g}] = [\underline{g}_1^*,\underline{g}_1^*]$ (central trick), condition (17) is equivalent to $[\underline{g}_1^*,\underline{g}_1^*] = \underline{g}_1^*$. The G-module $\underline{r}_u$ is generated by its 1-dimensional weight spaces $(\underline{r}_u)_\gamma$, γ degenerate sum (see Proposition 5.2). For each orbit of degenerate sums one $(\underline{r}_u)_\gamma$ suffices. We have $[(ds)\underline{g}_\alpha,(ds)\underline{g}_\beta] = [(\underline{g}_1^*)_\alpha,(\underline{g}_1^*)_\beta] \subset (\underline{r}_u)_{\alpha+\beta}$, because $d\phi$ is T-equivariant and $d\phi \circ ds = id$. (Use central trick.)

Hence we formulate the condition:

(18) For each orbit of degenerate sums, there is a pair of independent roots α,β, such that

1) $\alpha+\beta$ is in the orbit,

2) $[(ds)X_\alpha, (ds)X_\beta] \neq 0$.

It is clear that condition (18) is equivalent to (17). In the calculation of $[(ds)X_\alpha,(ds)X_\beta]$, we need a description of the composition $[\ ,\ ]$ on $\underline{g}_1^*$. The action of G on C induces one of $\underline{g}$ on C. In the Lie algebra $\ulcorner C,\underline{g}\urcorner$ of $\ulcorner C,G\urcorner$ we see from differentiation of Ad that

$[\ulcorner v,X\urcorner,\ulcorner w,Y\urcorner] = [X.w - Y.v, [X,Y]]$, for $X,Y \in \underline{g}$, $v,w \in C$.

(See [1], §3 for a similar situation.)

The Lie algebra $\underline{g}_1^*$ is a subalgebra of $\ulcorner C,\underline{g}\urcorner$, so

(19) $[(ds)X_\alpha,(ds)X_\beta] \neq 0$ if and only if $X_\alpha.\eta_3(df_1)X_\beta \neq X_\beta.\eta_3(df_1)X_\alpha$.

It follows from (9) that $(df_1)X_\alpha = \eta_2(dl_1)X_\alpha$. And $l_1(x_\alpha(t)) = x_\alpha(t).\eta_1 f_0 - \eta_1 f_0$. So $(dl_1)X_\alpha = X_\alpha.\eta_1 f_0$.

Summing up we get:

10.4 PROPOSITION.

<u>The sequences $\mathcal{E}_1,\mathcal{E}_2$ yield a k-extension ϕ as in Theorem 10.1 if and only if one of the following equivalent conditions is satisfied:</u>

(C1) $[\underline{g}_1^*,\underline{g}_1^*] = \underline{g}_1^*$,

(C2) $[\underline{g}_1^*,\underline{g}_1^*] \cap \underline{r}_u$ <u>generates</u> $\underline{r}_u$ <u>as a</u> G-<u>module</u>,

(C3) <u>For each orbit of degenerate sums, there is a pair of independent roots</u> α,β, <u>such that</u>

1) $\alpha+\beta$ <u>is in the orbit</u>,

2) $[(ds)X_\alpha,(ds)X_\beta] \neq 0$,

(C4) <u>For each orbit of degenerate sums, there is a pair of independent roots</u> α,β, <u>such that</u>

1) $\alpha+\beta$ <u>is in the orbit</u>,

2) $X_\alpha.(\eta_3\eta_2(X_\beta.\eta_1 f_0)) \neq X_\beta.(\eta_3\eta_2(X_\alpha.\eta_1 f_0))$.

10.5 The corresponding diagram is:

(All maps are T-equivariant, but the η_i are not G-equivariant).

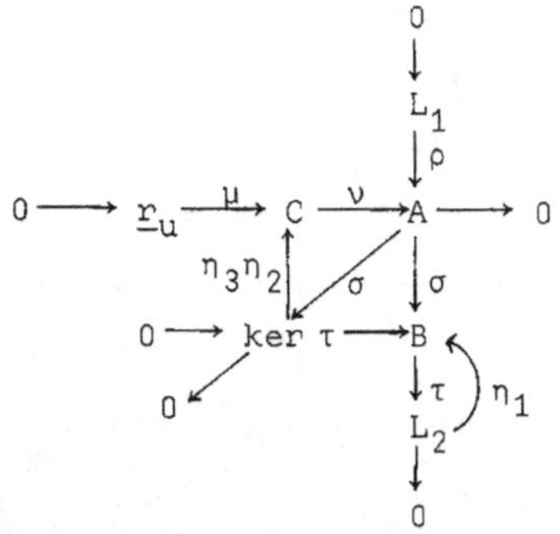

If one of the conditions (Ci) is satisfied, we say that condition (C) is satisfied.

10.6 (CASE BY CASE).

Now we have reached the point that we have to use the classification of degenerate sums. For each possible type we have to give $\mathcal{E}_1, \mathcal{E}_2$ satisfying condition (C). They have been found by trial and error. For non-exceptional types there is a non-trivial group Γ/Γ_0, which enables us to construct non-trivial 1-cocycles from lattices in $\underline{g}_{\mathbb{C}}$. For exceptional types we have to study other representations then the adjoint one. We will use the notations that are introduced in 4.

<u>A_2, characteristic 3.</u>

Root system $\{\alpha_1,\alpha_2,\alpha_1+\alpha_2, -\alpha_1,-\alpha_2,-\alpha_1-\alpha_2\}$.

Put $\alpha=\alpha_1$, $\beta=\alpha_2$, $\gamma=\alpha+\beta$. Let M_{st} denote the standard lattice in $\underline{g}_{\mathbb{C}}$, generated by X_β. It contains all X_δ and H_δ, $\delta \in \Sigma$. The G-module $A = L_{M_{st}}$ is isomorphic to $\underline{g}$ and contains an invariant 1-dimensional subspace L_1 generated by $\{H_\alpha+H_{-\beta}\}_{M_{st}}$. (It is the centre of $\underline{g}$, which is non-trivial because Γ/Γ_0 has p-torsion.) There is an admissable lattice M that is spanned by $\frac{1}{3}(H_\alpha+H_{-\beta})$ and M_{st}. (cf. 4.6). Let $\sigma : L_{M_{st}} \to L_M$ be the canonical homomorphism. Put $B = L_M$. Then dim A = dim B, ker $\sigma = L_1$, so $L_2 = B/\sigma A$ is 1-dimensional. (This is also clear from the fact that $L_2 = L_{M/M_{st}}$.) We get the exact sequence $\mathcal{E}_2 : 0 \to L_1 \xrightarrow{\rho} A \xrightarrow{\sigma} B \xrightarrow{\tau} L_2 \to 0$, where G acts trivially on L_1, L_2. As $A \simeq \underline{g}$, A fits in the exact sequence $\mathcal{E}_1 : 0 \to \underline{r}_u \to \underline{g}^* \to \underline{g} \to 0$.

Choose $f_0 = \{\frac{1}{3}(H_\alpha+H_{-\beta})\}_{M/M_{st}}$. We have to check condition (C) now. One has

$X_\alpha.(\eta_3\eta_2(X_\gamma.\eta_1 f_0)) = X_\alpha.(\eta_3\eta_2(X_\gamma.\{\frac{1}{3}(H_\alpha+H_{-\beta})\})) = 0$, and

$X_\gamma.(\eta_3\eta_2(X_\alpha.\{\frac{1}{3}(H_\alpha+H_{-\beta})\})) = -X_\gamma.(\eta_3\eta_2\{X_\alpha\}) = -\hat{a}d(X_\gamma)(X_\alpha^*) = \pm Z_{\alpha+\gamma}^* \neq 0$. (Use Proposition 3.3).

In the same way $[(ds)X_\gamma,(ds)X_\beta] = \ulcorner \pm Z_{\beta+\gamma}^*, 0 \urcorner \neq 0$. It is seen from 2.8 Table 1 that all orbits of degenerate sums are covered in this way.

10.7 REMARK.

One can avoid L_1 by dividing out the 1-dimensional submodules in $\underline{g}^*$ and $\underline{g}(=A)$. Then one doesn't need Theorem 9.6. In fact one returns to the following classical situation:

$0 \to \underline{r}_u \to C \to B \to L_2 \to 0$ is a resolution of L_2. In the same way the construction for D_1 and F_4 can be simplified. But it is not possible to do the same for B_1, G_2 in characteristic 2. At least not for the constructions that will be given below. In the case of G_2 in characteristic 3, we will use a construction where $L_1=0$ automatically. So we will need L_1 just in those cases that $\underline{r}_u$ has a 1-dimensional G-submodule (see Proposition 5.2).

Then we will use a sequence $\mathcal{E}_1$ in which A has a 1-dimensional G-submodule, which is the image of an indecomposable submodule of C that has dimension > 1.

10.8 A_3 and D_l, $l > 4$, characteristic 2.

We exploit the centre of $\underline{g}$ in the same way as above. The root system is $\Sigma = \{\pm\varepsilon_i \pm\varepsilon_j | 1 \leq i < j \leq l\}$. (See [4] "Planches" and use that $A_3 = D_3$). The element $X_{\varepsilon_1+\varepsilon_3}$ in $\underline{g}_{\mathbb{C}}$ generates a standard lattice M_{st}, corresponding to $\underline{g}$. (i.e. $L_{M_{st}} \simeq \underline{g}$). Choose $H = H_{\varepsilon_1+\varepsilon_1} + \sum_{i=1}^{l-1} H_{\varepsilon_i+\varepsilon_{i+1}}$. If l is even, then $H \in 2M_{st}$; if l is odd, then $\{H\}_{M_{st}}$ generates a 1-dimensional G-submodule. (It is the centre again). Anyway, $\frac{1}{2}H$ and M_{st} span an admissable lattice M'. In $L_{M'}$ the element $\{\frac{1}{2}H\}_{M'}$ generates a 1-dimensional G-submodule. So we can define the admissable lattice M, spanned by $\frac{1}{4}H$ and M_{st}. Let $\sigma : L_{M_{st}} \rightarrow L_M$ be the natural homomorphism, and choose

$$\mathcal{E}_2 : 0 \rightarrow L_1 \rightarrow L_{M_{st}} \rightarrow L_M \rightarrow L_2 \rightarrow 0.$$

Again we can identify $A = L_{M_{st}}$ with $\underline{g}$, and we put

$$\mathcal{E}_1 : 0 \rightarrow \underline{r}_u \rightarrow \underline{g}^* \rightarrow \underline{g} \rightarrow 0.$$

Choose $f_0 = \{\frac{1}{4}H\}$. (If l is even, then there is another factor of the centre. But that factor does not give the right cocycles). We check condition (C):

$$X_{\varepsilon_1+\varepsilon_2} \cdot (\eta_3\eta_2(X_{\varepsilon_1-\varepsilon_2}.\eta_1 f_0)) = 0,$$

$$X_{\varepsilon_1-\varepsilon_2} \cdot (\eta_3\eta_2(X_{\varepsilon_1+\varepsilon_2}.\eta_1 f_0)) = \bar{a}d(X_{\varepsilon_1-\varepsilon_2})(X^*_{\varepsilon_1+\varepsilon_2}) = Z^*_{2\varepsilon_1} \neq 0.$$

10.9 D_4, characteristic 2.

If we use the same construction as above, then it appears that one orbit of degenerated sums is missing:

$[(ds)X_{\varepsilon_1-\varepsilon_2},(ds)X_{\varepsilon_1+\varepsilon_2}] \neq 0$ and $[(ds)X_{\varepsilon_1-\varepsilon_2},(ds)X_{\varepsilon_3+\varepsilon_4}] \neq 0$,

but $[(ds)X_{\varepsilon_1+\varepsilon_2},(ds)X_{\varepsilon_3+\varepsilon_4}] = 0$.

So we have to do something about the orbit of $\varepsilon_1+\varepsilon_2+\varepsilon_3+\varepsilon_4$.
Say $\underline{r}_u = \underline{r}_1 \oplus \underline{r}_2 \oplus \underline{r}_3$, where $\underline{r}_1$ is the component that corresponds to the orbit of $2\varepsilon_1$, $\underline{r}_2$ to that of $\varepsilon_1-\varepsilon_2+\varepsilon_3+\varepsilon_4$, $\underline{r}_3$ to the last one.
What we have now is a 2-cocycle f_2 in $\underline{r}_u$ that behaves the right way in $\underline{r}_1 \oplus \underline{r}_2$. We divide out $\underline{r}_3$ and obtain a 2-cocycle f_2^{12} of G in $\underline{r}_1 \oplus \underline{r}_2$. We need a complementary cocycle in $\underline{r}_3$, to get a cocycle f_2^{123} in $\underline{r}_u$. From f_2^{12} we can get a 2-cocycle f_2^2 in $\underline{r}_1$ by dividing out $\underline{r}_2$. It is transformed into a suitable 2-cocycle in $\underline{r}_3$ by the automorphism of D_4 that interchanges the first and the third orbit.
We will use a slightly different method now. (It is not essentially different.) In $\underline{g}$ the element $H_{\varepsilon_1+\varepsilon_2} + H_{\varepsilon_1-\varepsilon_2}$ generates a 1-dimensional G-submodule. So we can choose the admissible lattice $M'' = \frac{1}{2}\mathbb{Z}(H_{\varepsilon_1+\varepsilon_2} + H_{\varepsilon_1-\varepsilon_2}) + M_{st}$ instead of the lattice $M = \mathbb{Z}(\frac{1}{4}H) + M_{st}$. Proceeding the same way as we did with M, we get a 2-cocycle for which $[(ds)\underline{g},(ds)\underline{g}] \cap \underline{r}_u = \underline{r}_2 \oplus \underline{r}_3$. Now we divide out the submodule $\underline{r}_1 \oplus \underline{r}_2$, and get a 2-cocycle f_2^3 in $\underline{r}_3$. The final 2-cocycle $f_2^{12} \oplus f_2^3$ in $\underline{r}_u$ satisfies condition (C).
Here one has to take for $\mathcal{E}_1$ the direct sum of
$0 \to \underline{r}_1 \oplus \underline{r}_2 \to \underline{g}^*/\underline{r}_3 \to \underline{g} \to 0$ and $0 \to \underline{r}_3 \to \underline{g}^*/\underline{r}_1 \oplus \underline{r}_2 \to \underline{g} \to 0$,
while $\mathcal{E}_2$ has to be the direct sum of the two corresponding $\mathcal{E}_2$'s.
Note that the sum of the cocycles that behave well in $\underline{r}_1 \oplus \underline{r}_2$ and $\underline{r}_2 \oplus \underline{r}_3$ respectively, is not behaving well in $\underline{r}_2$. That is the reason that $\underline{r}_2$ has to be divided out one time.

10.10 REMARK.

The reasoning we used for D_4 shows:
It is sufficient to construct for each orbit of degenerated sums a system $(\mathcal{E}_1,\mathcal{E}_2)$, such that there are α,β as in condition (C3) or (C4).

10.11 B_l, $l \geq 5$, characteristic 2.

We have seen earlier (in 3.11) how the Chevalley group G_{B_l} can be embedded in the Chevalley group $G_{D_{l+1}}$. From now on we will suppress the subscripts l in B_l and $l+1$ in D_{l+1}.

The embedding $G_B \to G_D$ induces a homomorphism $\underline{g}_B \to \underline{g}_D$, which in its turn induces a homomorphism of $\underline{g}_B^*$ into $\underline{g}_D^*$, given by

$X^*_{\varepsilon_i} \longmapsto X^*_{\varepsilon_i+\varepsilon_{l+1}} + X^*_{\varepsilon_i-\varepsilon_{l+1}}$ and $X^*_{\pm\varepsilon_i\pm\varepsilon_j} \longmapsto X^*_{\pm\varepsilon_i\pm\varepsilon_j}$.

The image of $(\underline{r}_u)_B$ in $\underline{g}_D^*$ is spanned by the elements $Z^*_{2\varepsilon_i}$ $(i \leq l)$, $Z^*_{2\varepsilon_{l+1}} + Z^*_{-2\varepsilon_{l+1}}$. We see that it has the same dimension as $(\underline{r}_u)_B$.

Hence there is an exact sequence of G_B-modules

$\mathcal{E}_1 : 0 \to (\underline{r}_u)_B \to (\underline{g}^*)_D \to A \to 0.$

As $(\underline{r}_u)_B$ is mapped into $(\underline{r}_u)_D$, there is a homomorphism

$A \to (\underline{g}^*)_D/(\underline{r}_u)_D \cong \underline{g}_D$.

Its kernel is 1-dimensional. (It is spanned by the image of $Z^*_{2\varepsilon_{l+1}}$.)

For D we used an exact sequence

$$0 \to L_1 \to \underline{g}_D \to B \to L_2 \to 0, \text{ where } \dim L_1 = 1.$$

Now we replace $\underline{g}_D \to B$ by $A \to B$, i.e. by the composite of $A \to \underline{g}_D$ and $\underline{g}_D \to B$, and get an exact sequence

$\mathcal{E}_2 : 0 \to L_1 \to A \to B \to L_2 \to 0$, where $\dim L_1 = 2$.

We have to check condition (C) again. For that purpose we may use the same calculation as we did for type D itself. It is also possible to calculate $[(ds)X_\alpha,(ds)X_\beta]$ using the fact that the Lie algebra of G_D^* is isomorphic to $\underline{g}_D^*$.

10.12 B_3, characteristic 2.

We still have an embedding $G_{B_l} \to G_{D_{l+1}}$ ($l = 3$ now). In the case of D_4 we did not use an exact sequence of the type

$0 \to (\underline{r}_u)_D \to \underline{g}_D^* \to \underline{g}_D \to 0$, but a direct sum of two sequences:

$0 \to \underline{r}_1 \oplus \underline{r}_2 \to \underline{g}_D^*/\underline{r}_3 \to \underline{g}_D \to 0$ and $0 \to \underline{r}_3 \to \underline{g}_D^*/\underline{r}_1 \oplus \underline{r}_2 \to \underline{g}_D \to 0$.
The image of $(\underline{r}_u)_B$ in $\underline{g}_D^*$ is spanned by the elements $Z^*_{2\varepsilon_i}$ (i=1,2,3), $Z^*_{2\varepsilon_4} + Z^*_{-2\varepsilon_4}$, $Z^*_{s_1\varepsilon_1+s_2\varepsilon_2+s_3\varepsilon_3+\varepsilon_4} + Z^*_{s_1\varepsilon_1+s_2\varepsilon_2+s_3\varepsilon_3-\varepsilon_4}$, $s_i = \pm 1$.
So $(\underline{r}_u)_B$ is mapped injectively into $\underline{g}_D^*/\underline{r}_3$. There is an exact sequence $\mathcal{E}_1 : 0 \to (\underline{r}_u)_B \to \underline{g}_D^*/\underline{r}_3 \to A \to 0$, and a natural homomorphism $A \to \underline{g}_D$, with 1-dimensional kernel. In the case of D_4 there was used an exact sequence $0 \to L_1 \to \underline{g}_D \to B \to L_2 \to 0$, corresponding to the sequence $0 \to \underline{r}_1 \oplus \underline{r}_2 \to \underline{g}_D^*/\underline{r}_3 \to \underline{g}_D \to 0$. Again we replace $\underline{g}_D \to B$ by $A \to B$, and we get an exact sequence $\mathcal{E}_2 : 0 \to L_1 \to A \to \to B \to L_2 \to 0$. It is easy to check condition (C) now.

REMARK 1.

We can't use the construction of case D_5 for the case B_4, because $(\underline{r}_u)_D$ is too small in this case. That is the reason that we will embed B_4 in F_4 and not in D_5.

REMARK 2.

For B_l, $l \geq 3$, $l \neq 4$, there also is a construction where dim A = dim $\underline{g}_{B_l}$. So this construction uses G-modules of lower dimension. (dim $\underline{g}_{B_l}$ < dim $\underline{g}_{D_{l+1}}$). In fact it uses a module A that is a quotient of the one used above.

10.13 <u>F_4, characteristic 2.</u>

We don't have a centre in $\underline{g}$ now, but we do have a G-submodule, generated by the X_α, α short. (See [26] Table 2).
It is spanned by the X_α, H_α, α short. (See also [22] page 155, Remark c.)
We put $M_{st} = \mathcal{U}_{\mathbb{Z}}(X_{\varepsilon_1+\varepsilon_2})$, $M_{\frac{1}{2}} = \mathcal{U}_{\mathbb{Z}}(\frac{1}{2}X_{\varepsilon_1})$. Then $M_{\frac{1}{2}} \supset M_{st}$.
There is a homomorphism of $G_{\mathbb{C}}$-modules, hence of $\mathcal{U}_{\mathbb{Z}}$-modules, $S : \underline{g}_{\mathbb{C}} \otimes \underline{g}_{\mathbb{C}} \to \underline{g}_{\mathbb{C}} \otimes \underline{g}_{\mathbb{C}}$ given by $S(x \otimes y) = x \otimes y + y \otimes x$.

Put

(2) $M' = 2M_{\frac{1}{2}} \otimes M_{st} \subset M_{st} \otimes M_{st}$,

(3) $M = 2M' + (M' \cap S(\mathfrak{g}_{\mathbb{C}} \otimes \mathfrak{g}_{\mathbb{C}}))$,

(4) $A = L_{M'/M}$.

Stated otherwise, A is the G-module that corresponds to the lattice M'_a that is the image of M' in $\mathfrak{g}_{\mathbb{C}} \wedge \mathfrak{g}_{\mathbb{C}} = \mathfrak{g}_{\mathbb{C}} \otimes \mathfrak{g}_{\mathbb{C}}/S(\mathfrak{g}_{\mathbb{C}} \otimes \mathfrak{g}_{\mathbb{C}})$.

Now we consider the element

(5) $H = H_\zeta \otimes H_{\varepsilon_1} + \sum_{\substack{\alpha \text{ short} \\ \alpha > 0}} X_\alpha \otimes X_{-\alpha}$, where $\zeta = \frac{1}{2}(\varepsilon_1+\varepsilon_2+\varepsilon_3+\varepsilon_4)$.

It is an element of M'.

We claim that $\{H\}_{M'/M}$ spans a 1-dimensional G-submodule in A.

Let H_a denote the image of H in $\mathfrak{g}_{\mathbb{C}} \wedge \mathfrak{g}_{\mathbb{C}}$. We have to prove that $\{H_a\}_{M'_a}$ spans a 1-dimensional G-submodule. First we prove that $\{H_a\}$ is invariant under W. It is clear that

(6) $\sum_{\substack{\alpha \text{ short} \\ \alpha > 0}} \{X_\alpha \wedge X_{-\alpha}\}$ is invariant under W.

Now we note that $H_{\varepsilon_1} \wedge H_{\varepsilon_1} = 0$ and

$(H_{\varepsilon_1} + H_{\varepsilon_i}) \wedge H_{\varepsilon_1} = 2H_{\varepsilon_1+\varepsilon_2} \wedge H_{\varepsilon_1} = -2H_{\varepsilon_1} \wedge H_{\varepsilon_1+\varepsilon_2} \in 2M'_a$ $(i \neq 1)$.

It follows that $\{H_{\varepsilon_i} \wedge H_{\varepsilon_1}\} = 0$ $(i \geq 1)$, whence

$\{H_\zeta \wedge H_{\varepsilon_1}\} = \{H_\alpha \wedge H_{\varepsilon_1}\}$, for α short, $(\alpha,\varepsilon_1) \neq 0$. (Inspect the root system). Using the action of W we see that

$\{H_\alpha \wedge H_\beta\} = \{H_\beta \wedge H_\alpha\} = \{H_\gamma \wedge H_\alpha\}$ if α,β,γ are short, $(\alpha,\beta) \neq 0$, $(\alpha,\gamma) \neq 0$.

Now let $w \in W$. Put $\alpha = w\zeta$, $\beta = w\varepsilon_1$. Then $(\alpha,\beta) \neq 0$. It follows from inspection of Σ that $(\alpha,\varepsilon_1) \neq 0$ or $(\beta,\varepsilon_1) \neq 0$. If $(\alpha,\varepsilon_1) \neq 0$, then $\{H_\alpha \wedge H_\beta\} = \{H_\alpha \wedge H_{\varepsilon_1}\} = \{H_\zeta \wedge H_{\varepsilon_1}\}$ and if $(\beta,\varepsilon_1) \neq 0$, then $\{H_\alpha \wedge H_\beta\} = \{H_{\varepsilon_1} \wedge H_\beta\} = \{H_\zeta \wedge H_{\varepsilon_1}\}$. We may conclude

(7) $\{H_a\}$ is invariant under W.

Now consider $x_{\varepsilon_2}(t)\{H_a\} - \{H_a\}$.

We have

$(x_{\varepsilon_2}(t)-1) \,.\, \{H_\zeta \wedge H_{\varepsilon_1} + X_{\varepsilon_2} \wedge X_{-\varepsilon_2}\} =$

$t\{-X_{\varepsilon_2} \wedge H_{\varepsilon_1} + X_{\varepsilon_2} \wedge H_{\varepsilon_2}\} = 2t\{X_{\varepsilon_2} \wedge H_{-\varepsilon_1+\varepsilon_2}\} = 0,$

$(x_{\varepsilon_2}(t)-1) \,.\, \{X_{\varepsilon_1} \wedge X_{-\varepsilon_1}\} = t\{2X_{\varepsilon_1+\varepsilon_2} \wedge X_{-\varepsilon_1}\} + t\{2X_{\varepsilon_1} \wedge X_{-\varepsilon_1+\varepsilon_2}\} +$

$+ 2t^2\{2X_{\varepsilon_1+\varepsilon_2} \wedge X_{-\varepsilon_1+\varepsilon_2}\} = 2t\{X_{-\varepsilon_1} \wedge X_{\varepsilon_1+\varepsilon_2}\} = 0,$

$(x_{\varepsilon_2}(t)-1) \,.\, (\{X_\zeta \wedge X_{-\zeta}\} + \{X_{\zeta-\varepsilon_2} \wedge X_{\varepsilon_2-\zeta}\}) = 2\{X_\zeta \wedge X_{\varepsilon_2-\zeta}\} = 0.$

All short roots that are orthogonal to ε_2 can be handled like ε_1. The remaining terms of H_a can be sorted in pairs of the type $\pm X_\gamma \wedge X_{-\gamma}$, $\pm X_{\gamma-\varepsilon_2} \wedge X_{\varepsilon_2-\gamma}$. They can be handled like the case $\gamma = \zeta$. It follows that

(8) $x_{\varepsilon_2}(t)$ fixes $\{H_a\}$.

Next consider $(x_{\varepsilon_2-\varepsilon_3}(t)-1).\{H_a\}$. Now we have

$(x_{\varepsilon_2-\varepsilon_3}(t)-1) \,.\, \{H_\zeta \wedge H_{\varepsilon_1}\} = 0,$

$(x_{\varepsilon_2-\varepsilon_3}(t)-1) \,.\, \{X_{\varepsilon_1} \wedge X_{-\varepsilon_1}\} = 0,$

$(x_{\varepsilon_2-\varepsilon_3}(t)-1) \,.\, (\{X_{\varepsilon_2} \wedge X_{-\varepsilon_2}\} + \{X_{\varepsilon_3} \wedge X_{-\varepsilon_3}\}) = 2t\{X_{\varepsilon_2} \wedge X_{-\varepsilon_3}\} = 0.$

Again all roots that are orthogonal to $\varepsilon_2-\varepsilon_3$ can be handled like ε_1, and again all remaining terms can be sorted in pairs $\pm X_\gamma \wedge X_{-\gamma}$, $X_{\gamma-\varepsilon_2+\varepsilon_3} \wedge X_{\varepsilon_2-\varepsilon_3-\gamma}$.

This finishes the proof of

(9) $\{H_a\}$ (or $\{H\}$) spans a 1-dimensional G-submodule in A. There is an admissible lattice in $\mathfrak{g}_{\mathbb{C}} \wedge \mathfrak{g}_{\mathbb{C}}$, spanned by $\frac{1}{2}H_a$ and M'_a. Let B denote the corresponding G-module, σ the natural map $A \to B$. We get $\mathcal{E}_2 : 0 \to L_1 \xrightarrow{\rho} A \xrightarrow{\sigma} B \xrightarrow{\tau} L_2 \to 0$.

Now we return to the first definition of A, $A = L_{M'/M}$, $M = 2M' + (M' \cap S(\mathfrak{g}_{\mathbb{C}} \otimes \mathfrak{g}_{\mathbb{C}}))$.

Put

(10) $M'' = 2M' + (M' \cap S(M_{st} \otimes M_{st}))$.

(11) $C = L_{M'/M''}$.

There is a natural map $\nu : C \to A$. We want to prove that there is an exact sequence of G-modules

$\mathcal{E}_1 : 0 \to \underline{r}_u \to C \xrightarrow{\nu} A \to 0$.

Hence consider ker ν. First we compare

$N'' = S(M_{st} \otimes M_{st}) \subset M_{st} \otimes M_{st}$ with $N = (M_{st} \otimes M_{st}) \cap S(\underline{g}_{\mathbb{C}} \otimes \underline{g}_{\mathbb{C}})$.

Choose a basis $e_1,\ldots,e_n$ of M_{st}. Then N'' is spanned by the elements

$e_i \otimes e_j + e_j \otimes e_i$ $(i \neq j)$, $2e_i \otimes e_i$.

And N is spanned by the elements

$e_i \otimes e_j + e_j \otimes e_i$ $(i \neq j)$, $e_i \otimes e_i$.

Now we specify the basis (e_i) of M_{st}, taking H_ζ, H_{ε_1}, $H_{\varepsilon_1-\varepsilon_2}$, $H_{\varepsilon_2-\varepsilon_3}$, X_α $(\alpha \in \Sigma)$.

We can obtain a basis of $M_{\frac{1}{2}}$ from it by dividing some of the e_i by 2. We reorder the basis in such a way that $\frac{1}{2}e_1, \frac{1}{2}e_2, \ldots$ $\ldots, \frac{1}{2}e_{26}, e_{27}, \ldots, e_{52}$ is a basis of $M_{\frac{1}{2}}$. Then M' is spanned by the elements $2e_i \otimes e_j$ $(i = 27,\ldots,52;\ j = 1,\ldots,52)$, $e_i \otimes e_j$ $(i = 1,\ldots,26;\ j = 1,\ldots,52)$.

Hence $M' \cap N$ differs from $M' \cap N''$ in the components spanned by the elements $e_i \otimes e_i$ $(i = 1,\ldots,26)$. It follows from $M = 2M' + (M' \cap N)$, $M'' = 2M' + (M' \cap N'')$ that ker ν is spanned by the elements $\{e_i \otimes e_i\}$, $i = 1,\ldots,26$. Note that $\{e_i \otimes e_i\}_{M'/M''} \neq 0$. It is clear that ker ν has dimension 26 and has a highest weight that is twice a short root. Then it follows from ([26], Table 2) that ker ν is irreducible.

From Proposition 5.2 we see that ker $\nu \simeq \underline{r}_u$. We have to check condition (C) now for $\mathcal{E}_1, \mathcal{E}_2$. Choose $f_0 = \{\frac{1}{2}H_a\}$.

We want to calculate

$$X_{\varepsilon_2+\varepsilon_3} \cdot (\eta_3\eta_2(X_{\varepsilon_2-\varepsilon_3} \cdot \eta_1 f_0)) - X_{\varepsilon_2-\varepsilon_3} \cdot (\eta_3\eta_2(X_{\varepsilon_2+\varepsilon_3} \cdot \eta_1 f_0)).$$

In order to do this, we fix the order on Σ:

For $a_i \in \mathbb{R}$ we define $a_1\varepsilon_1 + \ldots + a_4\varepsilon_4$ to be positive, if $a_1 = \ldots = a_{k-1} = 0$, $a_k > 0$ for some k, $1 \leqslant k \leqslant 4$. (This is the lexicographic order on $\mathbb{R}^4$).

Put $\Delta_1^+ = \{\alpha \in \Sigma | \alpha \text{ short}, \alpha+\varepsilon_2+\varepsilon_3 \in \Sigma,\ 2\alpha+\varepsilon_2+\varepsilon_3 > 0\}$,

$\Delta_1^- = \{\alpha \in \Sigma | \alpha \text{ short}, \alpha+\varepsilon_2+\varepsilon_3 \in \Sigma,\ 2\alpha+\varepsilon_2+\varepsilon_3 < 0\}$.

Define Δ_2^+, Δ_2^- in an analogous way, replacing $\varepsilon_2+\varepsilon_3$ by $\varepsilon_2-\varepsilon_3$.

Then we claim that

$$X_{\varepsilon_2+\varepsilon_3} \cdot (\eta_3\eta_2(X_{\varepsilon_2-\varepsilon_3} \cdot \eta_1 f_0)) =$$

$$X_{\varepsilon_2+\varepsilon_3} \cdot (\eta_3\eta_2\{\tfrac{1}{2}X_{\varepsilon_2-\varepsilon_3} \cdot (\sum_{\substack{\alpha>0\\ \alpha\in\Delta_2^+}} X_\alpha \wedge X_{-\alpha} - \sum_{\substack{\alpha>0\\ \alpha\in\Delta_2^-}} X_{-\alpha} \wedge X_\alpha + \sum_{\substack{\alpha<0\\ \alpha\in\Delta_2^-}} X_{-\alpha} \wedge X_\alpha$$

$$- \sum_{\substack{\alpha<0\\ \alpha\in\Delta_2^+}} X_\alpha \wedge X_{-\alpha})\}) = X_{\varepsilon_2+\varepsilon_3} \cdot \{\tfrac{1}{2}X_{\varepsilon_2-\varepsilon_3} \cdot (\sum_{\substack{\alpha>0\\ \alpha\in\Delta_2^+}} X_\alpha \otimes X_{-\alpha} -$$

$$- \sum_{\substack{\alpha>0\\ \alpha\in\Delta_2^-}} X_{-\alpha} \otimes X_\alpha + \sum_{\substack{\alpha<0\\ \alpha\in\Delta_2^-}} X_{-\alpha} \otimes X_\alpha - \sum_{\substack{\alpha<0\\ \alpha\in\Delta_2^+}} X_\alpha \otimes X_{-\alpha})\}_{M'/M''}.$$

Here the point is that the element Y inside $\{\ \}_{M'/M''}$ has to be in M'. This element Y is in the $\mathbb{C}$-span of the elements $X_\beta \otimes X_\gamma$, where β,γ are short roots with $\beta-\gamma > 0$.

The image in $\mathfrak{g}_{\mathbb{C}} \wedge \mathfrak{g}_{\mathbb{C}}$ is in the image M'_a of M'. It is easily derived from these facts (or from explicit calculation) that, indeed, $Y \in M'$. The element $X_{\varepsilon_2+\varepsilon_3}.Y$ of M' is a sum of terms

$\pm X_{\varepsilon_2+\varepsilon_3} \cdot \tfrac{1}{2}X_{\varepsilon_2-\varepsilon_3} \cdot (X_\alpha \otimes X_{-\alpha})$, α short.

For most roots α this term is zero. It is non-zero if

1) $\alpha + \varepsilon_2 + \varepsilon_3 + \varepsilon_2 - \varepsilon_3 = \alpha + 2\varepsilon_2 \in \Sigma$,

2) $\alpha + \varepsilon_2 + \varepsilon_3$ and $-\alpha + \varepsilon_2 - \varepsilon_3$ are in Σ,

3) Condition 1 or 2 holds for $-\alpha$ instead of α.

If condition 1 is fulfilled, then $\alpha = -\varepsilon_2$.

If condition 2 is fulfilled, then $\alpha+\varepsilon_2+\varepsilon_3$ is a short root such that $\alpha+\varepsilon_2+\varepsilon_3-2\varepsilon_2$ is a root, so $\alpha+\varepsilon_2+\varepsilon_3 = \varepsilon_2$.

We may conclude that $\alpha = \pm\varepsilon_2, \pm\varepsilon_3$ for non-vanishing terms of $X_{\varepsilon_2+\varepsilon_3}.Y$. Now it is easy to calculate $X_{\varepsilon_2+\varepsilon_3}.(\eta_3\eta_2(X_{\varepsilon_2-\varepsilon_3}.\eta_1 f_0))$.
It is

(12) $\{\frac{1}{2}X_{\varepsilon_2} \otimes [X_{\varepsilon_2+\varepsilon_3},[X_{\varepsilon_2-\varepsilon_3},X_{-\varepsilon_2}]] + \frac{1}{2}[X_{\varepsilon_2-\varepsilon_3},X_{\varepsilon_3}] \otimes [X_{\varepsilon_2+\varepsilon_3},X_{-\varepsilon_3}]\}$.

In the same way $X_{\varepsilon_2-\varepsilon_3}.\{\eta_3\eta_2(X_{\varepsilon_2+\varepsilon_3}.\eta_1 f_0)\} =$

$X_{\varepsilon_2-\varepsilon_3}.\{\frac{1}{2}X_{\varepsilon_2+\varepsilon_3}.(\sum_{\substack{\alpha>0 \\ \alpha\in\Delta_1^+}} X_\alpha \wedge X_{-\alpha} \;....)\} =$

$\{\frac{1}{2}X_{\varepsilon_2} \otimes [X_{\varepsilon_2-\varepsilon_3},[X_{\varepsilon_2+\varepsilon_3},X_{-\varepsilon_2}]] - \frac{1}{2}[X_{\varepsilon_2+\varepsilon_3},X_{-\varepsilon_3}] \otimes [X_{\varepsilon_2-\varepsilon_3},X_{\varepsilon_3}]\} =$

$\{\frac{1}{2}X_{\varepsilon_2} \otimes [X_{\varepsilon_2+\varepsilon_3},[X_{\varepsilon_2-\varepsilon_3},X_{-\varepsilon_2}]] - \frac{1}{2}[X_{\varepsilon_2-\varepsilon_3},X_{\varepsilon_3}] \otimes [X_{\varepsilon_2+\varepsilon_3},X_{-\varepsilon_3}]\}$.

Hence the difference with (12) is $\{X_{\varepsilon_2} \otimes X_{\varepsilon_2}\}$ which is non-zero.

10.14 B_4, characteristic 2.

There is a natural embedding of G_{B_4} in G_{F_4}, sending $x_{\pm\varepsilon_i}(t)$ to $x_{\pm\varepsilon_i}(t)$, and $x_{\pm\varepsilon_i\pm\varepsilon_j}(t)$ to $x_{\pm\varepsilon_i\pm\varepsilon_j}(t)$. (See 3.10, case 2 and 4.1 (1)).
We can exploit this embedding in exactly the same way as we exploited the embedding $G_{B_5} \to G_{D_6}$. We get

$\mathcal{E}_1 : 0 \to (\underline{r}_u)_{B_4} \to C_{F_4} \to A \to 0$ and

$\mathcal{E}_2 : 0 \to L_1 \to A \to B_{F_4} \to (L_2)_{F_4} \to 0$, where the subscript F_4 is used for modules that also occur in the construction for case F_4.

The dimension of L_1 is 2 again, and condition (C) is satisfied.

REMARK.

Every G_{B_4}-module is a direct sum of two components, one component containing all weight spaces with weights in Γ_0 (the lattice spanned by the roots), the other component containing other weight spaces (see Lemma 4.13). It follows that the system $\mathcal{E}_1, \mathcal{E}_2$ splits into two components. The component that corresponds to Γ_0 contains $\underline{r}_u$ and f_0. The other component may be deleted, which gives a construction with modules of lower dimensions.

10.15 G_2, characteristic 3.

We have the root system $\Sigma = \{\pm\alpha, \pm\beta, \pm\gamma, \pm(\alpha-\beta), \pm(\beta-\gamma), \pm(\gamma-\alpha)\}$, where $\alpha = -\alpha_1$, $\beta = 2\alpha_1 + \alpha_2$, $\gamma = -\alpha-\beta$.

We will need the signs of the structure constants $N_{\delta,\phi}$ $(\delta,\phi \in \Sigma)$.

It is possible to choose these signs in a "symmetric" way:

If r denotes a rotation of the root system over 60 degrees, then we require:

(1) $N_{r\delta,r\phi} = -N_{\delta,\phi}$. (see [22], p. 150).

We fix the signs by giving all X_δ in the 7-dimensional representation of $\mathfrak{g}_{\mathbb{C}}$:

$$(2)\quad X_\delta = \begin{pmatrix} & d & e & f & a & b & c \\ 2a & & g & h & & -f & e \\ 2b & j & & i & f & & -d \\ 2c & k & l & & -e & d & \\ 2d & & c & -b & & -j & -k \\ 2e & -c & & a & -g & & -l \\ 2f & b & -a & & -h & -i & \end{pmatrix},$$ where all variables except one are zero, one is 1.

(Empty entries are zero).

The variables correspond to the roots as indicated in the following illustration

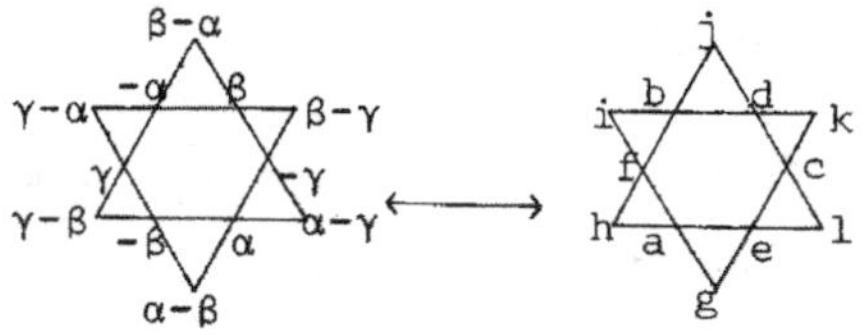

For example:

$$X_\alpha = \begin{pmatrix} & 0 & 1 & 0 & & & \\ & & & & & & 1 \\ & & & & & 0 & \\ & & & & -1 & & \\ 0 & & & & & & \\ 2 & & & & & & \\ 0 & & & & & & \end{pmatrix}, \quad X_{-\beta} = \begin{pmatrix} & & & & 1 & 0 & 0 \\ 2 & & & & & & \\ 0 & & & & & & \\ 0 & & & & & & \\ & 0 & & & & & \\ & & & 1 & & & \\ & & -1 & & & & \end{pmatrix}$$

It is seen that

(3) $N_{\alpha-\beta,\gamma-\alpha} = N_{\alpha-\beta,-\alpha} = N_{\beta,\alpha-\beta} = 1,\ N_{-\beta,\alpha} = 3,\ N_{\alpha,\gamma} = 2.$

Relation (1) follows from the fact that conjugation with the matrix $\begin{pmatrix} -1 & & & & & & \\ & & & & & 1 & \\ & & & & & & 1 \\ & & & & 1 & & \\ & & 1 & & & & \\ & & & 1 & & & \\ & 1 & & & & & \end{pmatrix}$ maps X_δ to $-X_{r\delta}$.

Relations (1), (3) determine all signs.

REMARK.

We will choose our definitions in such a way that their usefulness does not depend on the signs. But we need some choice of the signs to demonstrate their usefulness.

NOTATIONS.

Let (10) denote the highest weight of the 7-dimensional representation of $\underline{g}_{\mathbb{C}}$ and (01) the highest weight of the adjoint representation. Put (mn) = m(10) + n(01), R^{mn} = representation space of the irreducible representation of $\underline{g}_{\mathbb{C}}$ with highest weight (mn), $\overline{R}^{mn}$ = representation space of the irreducible G-module with highest weight (mn) (in characteristic p).

The $\underline{g}_{\mathbb{C}}$-module $R^{10} \otimes R^{01}$ has a direct summand of type R^{11}. The dimension of this factor is 64 (see [18], (5.9.4)).

The matrices X_δ in (2) are given with respect to some basis $e_1, e_2, \ldots, e_7$ in R^{10}.

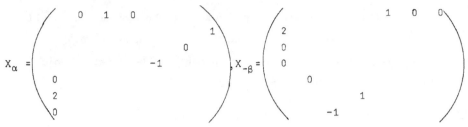

It is easily checked that $\{\sum_i n_i e_i \mid n_i \in \mathbf{Z}\}$ is an admissible lattice in R^{10}.

Let M_1 denote the standard lattice in $R^{01} = \mathfrak{g}_{\mathbb{C}}$, generated by $X_{\beta-\gamma}$. Put

(4) $v = e_5 \otimes X_{\beta-\gamma}$.

Then $v \in R^{11}_{2\beta-\gamma}$, because $e_5 \in R^{10}_{\beta}$. Put

(5) $M_{st} = \mathcal{U}_{\mathbf{Z}} v$.

Then M_{st} is a standard lattice in R^{11} that is contained in the admissible lattice

(6) $M = \{\sum_i e_i \otimes A_i \mid A_i \in M_1\}$ in $R^{10} \otimes R^{01}$.

We are interested in the G-module

(7) $R = L_{M_{st}+3M/3M}$

It is clear that R is a quotient of $L_{M_{st}}$. We claim that it is in fact isomorphic to $L_{M_{st}}$. After proving this claim, we will be able to recognize non-zero elements of $L_{M_{st}}$. The multiplicities of R^{11} are arranged like this:

```
          1   1
    1   2   2   2  (1)
  1   2   4   4   2   1
    2   4   4   4   2
  1   2   4   4   2   1
    1   2   2   2   1
          1   1
```

(see [21], Table 1.)

We use the same orientation as in the display of the root system, so the encircled multiplicity corresponds to the weight space of v.

Put

(8) $w = (X_{\alpha-\rho}X_{\gamma-\alpha} + X_{\gamma-\alpha}X_{\alpha-\beta})X_{\alpha}X_{\gamma}v$.

(We don't indicate the action by points now.)

This element w is in the weight space of weight 0.

$(w = e_1 \otimes H_\alpha - e_2 \otimes X_\beta + 2e_3 \otimes X_\alpha - e_4 \otimes X_\gamma - 2e_5 \otimes X_{-\beta} +$

$4e_6 \otimes X_{-\alpha} - 2e_7 \otimes X_{-\gamma}).$

In the following computations we put brackets around expressions that have the form $X_{\beta_1}\ldots X_{\beta_i}\ X_{\beta_{i+1}}\ldots X_{\beta_r}x$, where x is in M_{st} and $X_{\beta_{i+1}}\ldots X_{\beta_r}x$ is in a weight space that does not occur in R^{11}.

These expressions are zero, of course.

Put $x = X_\alpha X_\gamma v$.

Then $X_{\beta-\alpha}w = X_{\beta-\alpha}X_{\alpha-\beta}X_{\gamma-\alpha}x - (X_{\alpha-\beta}X_{\beta-\alpha}X_{\gamma-\alpha}x) + X_{\gamma-\alpha}X_{\beta-\alpha}X_{\alpha-\beta}x -$ $(X_{\gamma-\alpha}X_{\alpha-\beta}X_{\beta-\alpha}x) = H_{\beta-\alpha}X_{\gamma-\alpha}x + X_{\gamma-\alpha}H_{\beta-\alpha}x = 2X_{\gamma-\alpha}x + X_{\gamma-\alpha}x$. So

(9) $X_{\beta-\alpha}w = 3X_{\gamma-\alpha}x \in 3M$.

In the same way

(10) $X_{\alpha-\gamma}w = 3X_{\alpha-\beta}x \in 3M$.

So $[X_{\beta-\alpha}, X_{\alpha-\gamma}]w = 3X_{\beta-\alpha}X_{\alpha-\beta}x - (3X_{\alpha-\beta}X_{\beta-\alpha}x) - 3X_{\alpha-\gamma}X_{\gamma-\alpha}x +$ $(3X_{\gamma-\alpha}X_{\alpha-\gamma}x) = 3H_{\beta-\alpha}x - 3H_{\alpha-\gamma}x = 0$.

Hence

(11) $X_{\beta-\gamma}w = 0$.

Then $0 = X_{\gamma-\beta}H_{\beta-\gamma}w = X_{\gamma-\beta}X_{\beta-\gamma}X_{\gamma-\beta}w - (X_{\beta-\gamma}X^2_{\gamma-\beta}w) = H_{\beta-\gamma}X_{\gamma-\beta}w =$ $= -2X_{\gamma-\beta}w$, so

(12) $X_{\gamma-\beta}w = 0$.

Explicit calculation shows

(13) $X_\beta w \in 3M$.

It follows from (9)-(13) that $\underline{g}.\{w\} = 0$ in R. For $\alpha \in \Sigma$ we get $X_\alpha w \in 3M$, $(X_\alpha^2/2)w \in (\frac{3}{2}M) \cap M = 3M$, $(X_\alpha^3/6).w = 0$.

So $\mathcal{U}_{\mathbf{Z}}w \in 3M$, hence the non-zero element $\{w\}$ spans a 1-dimensional G-invariant subspace of R (see lemma 4.4). We want to find v_i that describe a composition series $v_1/v_2/\ldots/v_k$ in the sense of 4.14.

We have already found two of the composition factors: One is $\bar{R}^{11}$ with "generator" {v}, one is $\bar{R}^{00}$ with generator {w}.

We use the following table of multiplicities of weights in the $\bar{R}^{mn}$:

	(00)	(10)	(01)	(20)	(11)	(30)
(00)	1					
(10)	1	1				
(01)	1	0	1			
(20)	3	2	1	1		
(11)	1	3	1	2	1	
(30)	1	0	0	0	0	1

Table 2.

This table is obtained from ([21], Table 1,2).

In a row marked (mn) the multiplicities of the dominant weights in $\bar{R}^{mn}$ are given. These dominant weights are in the column headings. Using this table we will detect composition factors $\bar{R}^{01}$ and $\bar{R}^{10}$ of R. Then R has all composition factors of $L_{M_{st}}$, which proves the claim that $L_{M_{st}} \to R$ is an isomorphism. (The composition factors of $L_{M_{st}}$ are obtained from table 2 or from [21], Table 2).

Put

(14) $Y_1 = (X_\alpha X_\gamma + X_\gamma X_\alpha)v.$

This element is in the weight space of $\beta-\gamma$. One has $X_\beta Y_1 = 6v$, but $X_\beta X_{-\beta} v = 5v$. In $\bar{R}^{11}$ the weight $\beta-\gamma$ has multiplicity 1. We conclude that $\{X_{-\beta}v\}$ is mapped to a non-zero element and $\{Y_1\}$ is mapped to zero when the G-module spanned by {v} is mapped onto $\bar{R}^{11}$.

We express this fact by saying that $\{Y_1\}$ is zero in $\bar{R}^{11}$. (In fact $\bar{R}^{11} = L_{N'/N}$ for some N,N', and $\{Y_1\}_{N'/N} = 0$).

But $\{Y_1\}$ is non-zero in R.

So Y_1 corresponds to a composition factor $\overline{R}^{01}$ of R. Put

(15) $Y_2 = X_\gamma Y_1$.

Then $\{Y_2\}$ is non-zero in R, but $\overline{R}^{01}$ does not have weight β. So Y_2 corresponds to a factor $\overline{R}^{10}$. In summary, we get the composition series

(11) (01) (10) (00)

v / Y_1 / Y_2 / w .

Note that we don't claim that Y_2 generates w.

Put

(16) A = (R modulo the G-module generated by $\{Y_2\}$,$\{w\}$).

So A has composition factors $\overline{R}^{11}$, $\overline{R}^{01}$. Then A = $L_{M_{st}/N}$, N = = ker ($M_{st} \to A$). Let $M_{w/3}$ denote the lattice spanned by M_{st} and w/3 (Recall that $\{w\}$ is invariant in R $\cong L_{M_{st}}$.)

Put

(17) $B = L_{M_{w/3}/N}$.

The natural homomorphism $\sigma : A \to B$ is injective, because $\{w\} = 0$ in A. One obtains an exact sequence $\mathcal{E}_2 : 0 \to L_1 \to A \to B \to L_2 \to 0$, where $L_1 = 0$. Next we consider another representation of $\mathfrak{g}_{\mathbb{C}}$, in order to get the sequence $\mathcal{E}_1$. There is a homomorphism of $\mathfrak{g}_{\mathbb{C}}$-modules $[\ ,\ ] : \mathfrak{g}_{\mathbb{C}} \wedge \mathfrak{g}_{\mathbb{C}} \to \mathfrak{g}_{\mathbb{C}}$, defined by $[\ ,\ ]\ A \wedge B = [A,B]$. (Here $\mathfrak{g}_{\mathbb{C}} \wedge \mathfrak{g}_{\mathbb{C}}$ is the usual antisymmetric tensor product. See the case of F_4 above). The kernel of this homomorphism is R^{30}, as one sees from its dimension and its highest weight. We now proceed in $\mathfrak{g}_{\mathbb{C}} \wedge \mathfrak{g}_{\mathbb{C}}$, using only this factor R^{30} essentially. (In the same way as we only used R^{11} essentially in the construction of $\mathcal{E}_2$.)

In $\mathfrak{g}_{\mathbb{C}}$ we had the standard lattice M_1.

As the X_δ with δ short generate a proper G-submodule again (see table 2 and compare with the case of F_4 above), we can form the admissible lattice $M_{\frac{1}{3}}$, spanned by M_1 and the $\frac{1}{3}X_\delta$, $\frac{1}{3}H_\delta$. (δ short). (The submodule is an ideal of type $\bar{R}^{10}$.)

Consider $S = L_{M_1} \wedge M_{\frac{1}{3}}/3M_{\frac{1}{3}} \wedge M_{\frac{1}{3}}$.

It is easy to see that the multiplicities of weights in S give the following pattern:

$$\begin{array}{ccccccccccccc}
 & & & 1 & & 1 & & 1 & & 1 & & & \\
 & & 1 & & 2 & & 3 & & 2 & & \textcircled{1} & & \\
 & 1 & & 3 & & 3 & & 3 & & 3 & & 1 & \\
1 & & 2 & & 3 & & 4 & & 3 & & 2 & & 1 \\
 & 1 & & 3 & & 3 & & 3 & & 3 & & 1 & \\
 & & 1 & & 2 & & 3 & & 2 & & 1 & & \\
 & & & 1 & & 1 & & 1 & & 1 & & &
\end{array}$$

The weight $2\beta-\gamma$ has been marked by a circle again. Comparing with table 2, we conclude that S has composition factors $\bar{R}^{01}$, $\bar{R}^{01}$, $\bar{R}^{11}$, $\bar{R}^{30}$. (Two times $\bar{R}^{01}$.)

Put

(18) $v' = X_{\beta-\gamma} \wedge \frac{1}{3}X_\beta$.

This element corresponds to $\bar{R}^{11}$, because the other factors don't have the weight $2\beta-\gamma$.

Choose

(19) $Y_3 = (X_\alpha X_\gamma + X_\gamma X_\alpha)v'$,

(20) $Y_4 = X_{\beta-\alpha}Y_3$,

(21) $Y_5 = X_{\alpha-\beta}Y_4$.

Calculation shows that $\{Y_3\}$, $\{Y_4\}$, $\{Y_5\}$ are non-zero in S. As Y_1 was zero in $\bar{R}^{11}$ in the case of R, the element Y_3 is zero in $\bar{R}^{11}$ now. (They have the same image in $\bar{R}^{11}$.) So Y_3 corresponds to a factor $\bar{R}^{01}$ in S. Y_4 corresponds to a factor $\bar{R}^{30}$, because its image in $\bar{R}^{11}$ (from Y_3) is zero (see table 2). Then Y_5 corresponds to a factor $\bar{R}^{01}$, because its image in $\bar{R}^{30}$ is zero (see table 2). We get

(11) (01) (30) (01)

v' / Y_3 / Y_4 / Y_5 .

Put

(22) C = (S modulo the G-module generated by $\{Y_5\}$),

(23) A'= (S modulo the G-module generated by $\{Y_4\}$).

(Note that $\{Y_4\}$ generates $\{Y_5\}$).

This gives $\mathcal{E}_1 : 0 \to \underline{r}_u \to C \xrightarrow{\nu} A' \to 0$.

($\underline{r}_u$ is of type $\overline{R}^{30}$, so $\underline{r}_u \simeq \ker \nu$).

We have to prove that A $\simeq$ A', before we can check condition (C). Both A and A' have composition factors $\overline{R}^{11}$, $\overline{R}^{01}$. Furthermore they have composition series

(11) (01)

v / $(X_\gamma X_\alpha + X_\alpha X_\gamma)v$

and (11) (01)

v' / $(X_\gamma X_\alpha + X_\alpha X_\gamma)v'$ respectively.

We prove from these facts that A $\simeq$ A'. The proof closely resembles the proof for irreducible G-modules (see [2], 5.3).

In the G-module A $\oplus$ A' we choose the G-submodule A" generated by $\{v\} \oplus \{v'\} \in A \oplus A'$. As the G-submodules are the $\mathcal{U}_{\mathbb{Z}}$-submodules (see Lemma 4.4), the elements

$$\frac{X_{\beta_1}^{n_1}}{n_1!} \cdots \frac{X_{\beta_k}^{n_k}}{n_k!} \binom{H_{\alpha_1}}{m_1} \cdots \binom{H_{\alpha_l}}{m_l} \frac{X_{\beta_{k+1}}^{n_{k+1}}}{n_{k+1}!} \cdots \frac{X_{\beta_{2k}}^{n_{2k}}}{n_{2k}!} (\{v\} \oplus \{v'\})$$

where $\beta_1 < \ldots < \beta_{2k}$ are the roots, α_i are the simple roots, span A (see [22], Theorem 2). We have $X_{\gamma-\alpha}(X_\alpha^2/2)(\{v\} \oplus \{v'\}) = 0$. The element $\{v\} \oplus \{v'\}$ is a highest weight vector, so the weight space $A''_{\beta-\gamma}$ is spanned by $X_{-\beta}(\{v\} \oplus \{v'\})$ and $X_\gamma X_\alpha(\{v\} \oplus \{v'\})$. (Note that $-\beta < \gamma < \alpha < 0$ in the ordering that makes $2\beta-\gamma$ dominant.)

The weight spaces of $\beta-\gamma$ in A and A' are spanned by the corresponding elements. So the kernel of the projection of A" on A (or A') has no weight $\beta-\gamma$. (The image of $A''_{\beta-\gamma}$ has dim. 2 as one sees from table 2 and these remarks.) Then it has no kernel at all, because all composition factors of $A \oplus A'$ have $\beta-\gamma$ as a weight. We conclude that $A \simeq A'' \simeq A'$.

Now we can check condition (C).

Choose $f_0 = \{\frac{1}{3}w\}$. Then $X_{\beta-\alpha} \cdot (\eta_3\eta_2(X_{\beta-\gamma}\cdot\eta_1 f_0)) = 0$ (See (11).)

And $X_{\beta-\gamma} \cdot (\eta_3\eta_2(X_{\beta-\alpha}\cdot\eta_1 f_0)) = X_{\beta-\gamma}(\eta_3\eta_2\{X_{\gamma-\alpha}X_\alpha X_\gamma v\}) =$ $\{X_{\beta-\gamma}X_{\gamma-\alpha}X_\alpha X_\gamma v'\}$. (See (9).)

This element is non-zero in C. (It is $\{-2X_{\beta-\gamma} \wedge X_{\beta-\alpha}\}$, which is non-zero in S.)

10.16 <u>G_2, characteristic 2.</u>

We use the same kind of notations R^{mn}, $\overline{R}^{mn}$ as above. (But p = 2 for $\overline{R}^{mn}$, of course.)

In R^{10} we use the same basis $e_1,\ldots,e_7$. Put $M_{st} = \{\sum_i n_i e_i \mid n_i \in \mathbb{Z}\}$.

In $L_{M_{st}}$ there is a 1-dimensional G-submodule, spanned by $\{e_1\}$.

So we can form the admissible lattice $M_{\frac{1}{2}}$ spanned by $\frac{1}{2}e_1$ and M_{st}.

We need a table like table 2, but for p = 2. It is the table

	(00)	(10)	(01)	(20)
(00)	1			
(10)	0	1		
(01)	2	1	1	
(20)	0	0	0	1

Table 3.

The multiplicities of $\overline{R}^{00}$, $\overline{R}^{10}$, $\overline{R}^{01}$ are calculated by hand and those for $\overline{R}^{20}$ then follow from the Steinberg Tensor Product Theorem (see [22], p. 217). Note that $\overline{R}^{10} = L_{M_{st}/2M_{\frac{1}{2}}}$, and that the table says that $\underline{g}$ has no proper invariant subspaces. (It has no centre because $\Gamma = \Gamma_0$ and furthermore X_δ generates $\underline{g}$ for all roots δ.)

In $R^{10} \otimes R^{10}$ we have the lattices $M_{\frac{1}{2}} \otimes M_{st}$ and $2M_{\frac{1}{2}} \otimes M_{\frac{1}{2}}$, the former containing the latter. Put

(1) $S = L_{M_{\frac{1}{2}} \otimes M_{st}/2M_{\frac{1}{2}} \otimes M_{\frac{1}{2}}}$.

It has multiplicities 6, 3, 2, 1, hence composition factors $\overline{R}^{20}$, $\overline{R}^{01}$, $\overline{R}^{01}$, $\overline{R}^{10}$, $\overline{R}^{00}$, $\overline{R}^{00}$. Choose

(2) $Y_1 = e_4 \otimes e_5$, $Y_2 = X_{-\alpha}Y_1$, $Y_3 = \frac{X^2_{-\beta}}{2} Y_2$, $Y_4 = X_{\beta-\gamma}Y_3$.

Note that $\{Y_1\} \in S_{\beta-\gamma}$.

Calculation shows that $\{Y_i\} \neq 0$ in S. The submodule generated by $\{Y_1\}$ has at least the composition factors $\overline{R}^{01}$ (from Y_1), $\overline{R}^{20}$ (from Y_2), $\overline{R}^{01}$ (from Y_4). This is seen from table 3 in the same way as above (see 10.15).

If one divides out in S the G-module generated by $\{Y_1\}$, then the result has a factor $L_{M_{\frac{1}{2}} \otimes M_{st}/M_{st} \otimes M_{st} + 2M_{\frac{1}{2}} \otimes M_{\frac{1}{2}}}$. This module is of type $\overline{R}^{10}$ with generator $\{Z_1\}$ where

(2) $Z_1 = \frac{e_1}{2} \otimes e_5 + e_3 \otimes e_4$.

(Use multiplicities again).

There are composition factors $\overline{R}^{00}$ missing still. One of them corresponds to Y_3. Proof:

Let S_4 denote the G-module generated by $\{Y_4\}$. As we know multiplicities outside weight zero, we can check that S_4 has the following base outside weight zero:

$\{\{e_i \otimes e_j + e_j \otimes e_i\} | 1 \leq i < j \leq 7,\ j \neq i+3 \text{ or } j = 4\}$.

The weight space $(S_4)_0$ is spanned by the images of this base under the action of the X_δ. One checks that $(S_4)_0$ has the base

$\{e_2 \otimes e_5 + e_5 \otimes e_2 + e_4 \otimes e_7 + e_7 \otimes e_4\}$,

$\{e_2 \otimes e_5 + e_5 \otimes e_2 + e_3 \otimes e_6 + e_6 \otimes e_3\}$, which does not span $\{Y_3\}$.

Only one composition factor is missing now. But we want to do more than finding this last one: We want to change the order in the composition series. (Y_1 is following Z_1 now, but we want Z_1 to follow Y_1).

So consider S/S_2, where S_2 is generated by $\{Y_2\}$. We know that $\{Y_1\}$ generates a G-submodule in S/S_2 that has no composition factor $\bar{R}^{10}$. In S/S_2 one checks that $X_\delta\{Z_1\} = 0$ for δ positive, and that $\frac{X_\delta^n}{n!}\{Z_1\} = 0$ for $n > 1$. (The last result is obtained from the multiplicities.) Hence $\{Z_1\}$ is a highest weight vector of a G-submodule of S/S_2. (Use the standard base of $\mathcal{U}_{\mathbb{Z}}$ as in the case $p = 3$.)

Then $\{Z_1\}$ generates a G-submodule of S/S_2 without composition factor $\bar{R}^{01}$. Choose

(4) $Z_2 = X_{-\beta}Z_1$.

Check that $\frac{X_\beta^2}{2}\{Z_2\} = \{Y_2\}$ in S.

The conclusion is that S has the composition series

(01) (10) (00) (20) (00) (01)

Y_1 / Z_1 / Z_2 / Y_2 / Y_3 / Y_4 .

Choose

C = the G-module generated by $\{Z_1\}$ modulo the G-module generated by $\{Y_4\}$. It has composition series

(10) (00) (20) (00)

Z_1 / Z_2 / Y_2 / Y_3 .

In C the element Y_2 generates a G-module isomorphic to $\underline{r}_u$. (This is seen as in the proof of Proposition 5.2.) Hence one gets

$\mathcal{E}_1 : 0 \to \underline{r}_u \to C \to A \to 0.$

Then $A \simeq L_{M_{st}}$. Choose $B = L_{M_{\frac{1}{2}}}$. That gives

$\mathcal{E}_2 : 0 \to L_1 \to A \to B \to L_2 \to 0.$

We check condition (C):

$X_\beta \cdot (\eta_3\eta_2(X_{\alpha-\gamma}\cdot\eta_1 f_0)) = 0$, but

$X_{\alpha-\gamma} \cdot (n_3 n_2(X_\beta . n_1 f_0)) = X_{\alpha-\gamma} \cdot (n_3 n_2\{e_5\}) = \{X_{\alpha-\gamma} Z_1\} = \{e_3 \otimes e_3\} \neq 0.$

End of Proof of Theorem 10.1.

10.17 REMARK.

In the case by case part one may use more embeddings of Chevalley groups in Chevalley groups to get proofs like that for type B_1. There are useful embeddings $A_2 \to G_2$ ($p = 3$), $D_4 \to B_4$ ($p = 2$), $G_2 \to D_4$ ($p = 2$). The last one corresponds to the fixed points of the triality automorphism of D_4, and can be described analogously to 3.11. In this case one has to divide out two of the $\underline{r}_i$ in $(\underline{r}_u)_{D_4} = \underline{r}_1 \oplus \underline{r}_2 \oplus \underline{r}_3$, in order to get a close resemblance of $(\underline{r}_u)_{G_2}$ and the G_{D_4}-module. (cf. case B_3. See 10.12.) The modules C in these alternative proofs have higher dimensions.

10.18 Let $\phi : G^* \to G$ be the extension that is constructed in the proof of Theorem 10.1. So G^* is a subgroup of $\ulcorner C,G \urcorner$, containing $\ulcorner \underline{r}_u,1 \urcorner$, where C is a G-module containing $\underline{r}_u$ (see 10.3 (13)). The map ϕ is the restriction of $p_G : \ulcorner C,G \urcorner \to G$ to G^*. We may and shall assume that all weights of C are in Γ_0 (see 10.14 Remark). Let G_{ad} denote the adjoint group corresponding to G. Then there is a natural homomorphism $\ulcorner id,Ad \urcorner : \ulcorner C,G \urcorner \to$ $\to \ulcorner C,G_{ad} \urcorner$.

10.19 DEFINITION.

The image of G^* under $\ulcorner id,Ad \urcorner$ is denoted G^*_{ad} and the restriction of $p_{G_{ad}} : \ulcorner C,G_{ad} \urcorner \to G_{ad}$ to G^*_{ad} is denoted ϕ_{ad}. So $\phi_{ad} : G^*_{ad} \to G_{ad}$ is an extension of G_{ad} by $\underline{r}_u$.

10.20 PROPOSITION.

Assume $\Sigma \cap p\Gamma = \emptyset$ and $\underline{r}_u \neq 0$.

The extension ϕ_{ad} is non-trivial, i.e. there is no homomorphism $s : G_{ad} \to G^*_{ad}$ satisfying $\phi_{ad} \circ s = id$.

PROOF.

Suppose s exists. Put $\chi = \ulcorner id, Ad \urcorner$. Then $\phi_{ad} \circ \chi = Ad \circ \phi$, hence $d\phi_{ad} \circ d\chi = ad \circ d\phi$. Consider the inverse image $\underline{h}$ of $(ds)\underline{g}_{ad}$ in $\underline{g}^*$. It is a Lie algebra. If $\gamma \in \Sigma$, then $X^*_\gamma + \underline{r}_u$ is mapped onto the inverse image of $ad(X_\gamma)$ in $\underline{g}^*_{ad}$, so $ad(X_\gamma)$ is contained in $(d\phi_{ad} \circ d\chi)(\underline{h})$. It follows from the central trick that $\underline{h}$ contains all $[X^*_\alpha, X^*_\beta]$, with $\alpha, \beta \in \Sigma$. Hence it contains non-trivial elements of $\underline{r}_u$ (see Theorem 3.5 and Corollary 3.14). But then $(ds)\underline{g}_{ad}$ contains non-trivial elements of $\ulcorner \underline{r}_u, 0 \urcorner = \ker(d\phi_{ad})$, which contradicts $\phi_{ad} \circ s = id$.

10.21 Let G^* be contained in $\ulcorner C, G \urcorner$ as above. Let $N_{\ulcorner C,G \urcorner} G^*$ ($Z_{\ulcorner C,G \urcorner} G^*$) denote the normalizer (centralizer) of G^* in $\ulcorner C, G \urcorner$. Then $N_{\ulcorner C,G \urcorner} G^* / Z_{\ulcorner C,G \urcorner} G^*$ acts faithfully on G^*. We will see later (in 13.7) that $Int(N_{\ulcorner C,G \urcorner} G^*)$ is a subgroup of finite index in $Aut(G^*) = \{\psi \mid \psi$ is an automorphism of G^* in the category of algebraic groups$\}$. At this moment we only prove:

10.22 PROPOSITION.

Let $\underline{r}_u$ be non-zero. Then

$\dim (N_{\ulcorner C,G \urcorner} G^* / Z_{\ulcorner C,G \urcorner} G^*) \geqslant \dim G^*$.

PROOF.

The proof is easy if the centre $Z(G^*)$ of G^* has dimension zero. So we assume that the connected component of $Z(G^*)$ is non-trivial.

Then it corresponds to the 1-dimensional G-submodule of $\underline{r}_u$ and Σ is of type B_1 or G_2, $p = 2$ (see Proposition 5.2). So $Z(G^*)$ is 1-dimensional. Inspection of the constructions in 10.11, 10.12, 10.14, 10.16 shows (cf. Remark 10.7) that the inverse image of L_1 in C is a G-submodule that contains $\underline{r}_u$ as a submodule but not as a direct factor. The elements of this inverse image are mapped into $N_{\ulcorner C,G \urcorner} G^*$ by $i_C : C \to \ulcorner C,G \urcorner$, but some of them are not mapped into $R_u \cdot Z_{\ulcorner C,G \urcorner} G^*$ (otherwise $\underline{r}_u$ would be a direct factor). Hence $\dim (N_{\ulcorner C,G \urcorner} G^*/Z_{\ulcorner C,G \urcorner} G^*) >$ $\dim (G^*/Z(G^*)) \geqslant \dim (G^*)-1$.

10.23 REMARK.

There is a natural representation of $\ulcorner C,G \urcorner$ and hence of G^*. Its representation space is $K \oplus C$ and its action is defined by $\ulcorner v,g \urcorner.(\xi,v') = (\xi,\xi v+g.v')$. If we assume as in 10.18 that all weights of C are in Γ_0, then the image of the representation is isomorphic to G^*_{ad}. If, on the contrary, we replace C by a bigger representation (adding direct summands for instance) such that the weights span Γ, then the image is isomorphic to G^*. Intermediate lattices of weights yield "intermediate" images.

10.24 REMARK.

The irreducible (rational) representations of G^* correspond to the irreducible rational representations of G, because the fixed points of R_u constitute an invariant subspace.

10.25 REMARK.

Let s be a cross section of ϕ as in Theorem 8.2 (cf. 10.3, (14)). Then $\pi \circ Ad_{G^*}(sx) = Ad_G(\phi(sx)) \circ \pi = Ad_G(x) \circ \pi$, and hence $Ad_{G^*} \circ s = \hat{A}d$.

It follows that $\hat{a}d = ad \circ ds$, or $\hat{a}d \circ \pi = ad$, which was proved in 3.3.

10.26. REMARK.
If all roots are long then the adjoint representation of G^*_{ad} is faithful. It then induces a representation of G^* that is isomorphic to the representation obtained from 10.7, 10.23.

§11. Relations in the open cell.

In this section we consider an arbitrary solution $\phi : G^* \to G$ of the problem $d\phi = \pi$ (see section 7). Fixing a maximal torus T^* in G^*, we derive relations between elements in T^*-stable unipotent subgroups of G^*. These relations are the analogues of relations (A), (B) in Steinbergs set of defining relations for G (see [23] or [22], §6). As a result of these relations we will show that ker ϕ is abelian in most cases (see 11.21).

11.1 Let $\phi : H \to G$ be a surjective separable k-homomorphism of connected algebraic groups, where G is an almost simple Chevalley group with $[\underline{g},\underline{g}] = \underline{g}$. Let $\underline{h}$ denote the Lie algebra of H, T the usual maximal torus in G, T^* a k-torus in H satisfying $\phi T^* = T$. Assume that T^* is k-split.
If G is simply connected, let $\underline{r}_u$ be the G-module described in 5.2 (cf. section 10). If not, put $\underline{r}_u = 0$ (cf. Lemma 7.1). In both cases $\underline{r}_u$ can be viewed as an H-module by means of ϕ.
Now we introduce three properties (arranged in order of increasing strength).

(P1) <u>There is a homomorphism of</u> H-<u>modules</u> $\mu : \underline{r}_u \to \ker(d\phi)$ <u>such that</u> T^* <u>acts trivially on the cokernel of</u> μ.

(P2) <u>There is an</u> H-<u>equivariant</u> k-<u>homomorphism</u> τ <u>from</u> $\underline{r}_u$ <u>into</u> ker ϕ

such that $d\tau = \mu$ *is as in* (P1).

(P3) $\tau(\underline{r}_u) = \ker \phi$, *where* τ *is as in* (P2).

REMARKS.

1) In (P2) it is sufficient to assume that τ maps into H, because $\phi \circ \tau(\underline{r}_u)$ is a connected unipotent normal subgroup.

2) If (P1) holds, then $\mu(\underline{r}_u)$ is contained in the Lie algebra of $R_u(H)$. Proof: Consider the natural projection $\psi : \ker \phi \to \ker \phi/R_u(H)$. As $\ker \phi$ acts trivially on $\mu(\underline{r}_u)$, a maximal torus of $\ker \phi/R_u(H)$ acts trivially on $(d\psi \circ \mu)(\underline{r}_u)$. It follows from ([1], Theorem (13.18)) that $(d\psi \circ \mu)(\underline{r}_u)$ consists of semi-simple elements. On the other hand it follows as in 6.2 that $(d\psi \circ \mu)(Z_\gamma^*)$ is nilpotent for γ degenerate. These elements $(d\psi \circ \mu)(Z_\gamma^*)$ generate $(d\psi \circ \mu)(\underline{r}_u)$ (see Proposition 5.2), so $(d\psi \circ \mu)(\underline{r}_u) = 0$.

EXAMPLES.

1) If $\phi : G^* \to G$ is a solution of $d\phi = \pi$, as described in 7.2, then ϕ satisfies (P1). We will see in 11.21, 11.27 that ϕ also satisfies (P3), with one possible exception.

2) If $\phi : G^* \to G$ is the extension of G by $\underline{r}_u$, constructed in section 10, then ϕ satisfies (P3). (Then it also satisfies (P1) (P2), of course.)

3) If $\phi = p_G : \ulcorner \underline{r}_u, G \urcorner \to G$ (see 8.1), then ϕ also satisfies (P3).

11.2 LEMMA.

If $\phi : H \to G$ *satisfies* (P2), *then* $Ad_H \circ \tau$ *is trivial (i.e. it maps* $\underline{r}_u$ *to* 1).

PROOF.

Let $X \in \underline{r}_u$. For $x \in H$ we have $(x,\tau(X)) = \tau(x.X-X)$, so the morphism $x \mapsto (x,\tau(X))$ has differential zero (see Proposition 5.2 and use that $dFr = 0$). But this differential is also equal to $id-Ad_H(\tau(X))$ (see [1], (3.9)).

11.3 In the sequel we shall derive several results about $\phi : G^* \to G$ which do not depend on the property $d\phi = \pi$, but only on (P1), (P2) or (P3). We shall apply those results in situations like example 3 in 11.1. Therefore we shall label such results with the corresponding properties, suppressing (P1) if (P2) holds and (P2) if (P3) holds. So a label (P1) means that some natural modifications yield a result that is valid if (P1) holds in 11.1. (It doesn't mean that (P1) is necessary.) We give some examples of these modifications:

Replace G^* by H, replace Z^*_γ by $\mu(Z^*_\gamma)$, if necessary.

Replace X^*_α by the weight vector of T^* in $\underline{h}$ that satisfies $(d\phi)X^*_\alpha = X_\alpha$. Omit weights that don't occur in $\underline{h}$.

We shall give proofs of labeled statements only for the case $\phi : G^* \to G$, leaving the general case to the reader.

11.4 We return to $\phi : G^* \to G$ with $d\phi = \pi$ (see 7.2). Assume that ϕ is defined over k and that $\underline{r}_u \neq 0$. So G is simply connected almost simple, $\Sigma \cap p\Gamma = \emptyset$ (see Proposition 1.3 (ii) and Proposition 2.2) and Γ contains degenerate sums (see 3.14). We know that $\ker\phi$ is the unipotent radical R_u of G^* (see Lemma 7.4). It follows from ([1], (6.7) Remark) that R_u is defined over k. The inverse image $\phi^{-1}(T)$ of T is also defined over k (see [1], (6.7), (6.8), applied to the action of G^* on G/T). Hence $\phi^{-1}(T)$ contains a maximal torus T^*, defined over k (see [1], (18.2)). This torus T^* is mapped isomorphically onto T. So T^* is k-split.

(Note that this was assumed in 11.1.) The action Ad of G^* on $\underline{g}^*$ is given by $Ad(x)(X) = \hat{A}d(\phi x)(X)$ for $x \in G^*$, $X \in \underline{g}^*$ (see 7.2). So the weight spaces of $Ad : G^* \to \underline{g}^*$ are the same as those of $\hat{A}d$ (for T^*, T respectively). Henceforth we identify weights on T^* with weights on T.

REMARK.

In the following Proposition short roots have to be handled with special care, because a p-multiple of a short root is a degenerate sum (see Lemma 2.9, (iii)).

11.5 PROPOSITION (P1).(cf. [7], Exp. 13, Th. 1).

Let γ be a non-zero weight of $\underline{g}^*$.

(i) If γ is not a short root, then there is a connected subgroup G^*_γ of G^*, defined over k, such that

(a) The Lie algebra of G^*_γ is $\underline{g}^*_\gamma$.

(b) As an algebraic group, G^*_γ is T^*-equivariantly k-isomorphic to $\underline{g}^*_\gamma$.

(ii) If γ is a short root, then there is a T^*-equivariant k-isomorphism of varieties from $\underline{g}^*_\gamma$ into G^*, mapping 0 to 1.

PROOF.

(i) The multiplicity of γ is 1, and the multiplicity of $n\gamma$ is zero for $n > 1$ (use Lemma 2.6 (i) and Proposition 2.12). So it follows from ([3], Theorem 9.16), that there is a T^*-stable subgroup G^*_γ satisfying (a). It is the unipotent radical of $T^*G^*_\gamma$. Now (b) follows from ([3], Theorem 9.8).

(ii) The multiplicity of γ is 1, the multiplicity of $p\gamma$ is 1 and those of other positive multiples of γ are zero (see Lemma 2.9 (iii), Lemma 2.6 (i), Proposition 5.2). Hence we get from ([3], Theorem 9.16) the existence of a connected T^*-stable subgroup $G^*_{(\gamma)}$ of G^* with Lie algebra $\underline{g}^*_\gamma$ or $\underline{g}^*_\gamma + \underline{g}^*_{p\gamma}$. Again $G^*_{(\gamma)}$ is the unipotent radical of $T^*G^*_{(\gamma)}$. The centralizer of T^* has trivial intersection with $G^*_{(\gamma)}$ (see [1], Proposition 9.4), so (ii) follows from ([3], Corollary 9.12).

11.6 Let γ be a weight as in Proposition 11.5, (i). We identify the additive group $\underline{g}^*_\gamma$ with its Lie algebra. Then the isomorphism $\theta : \underline{g}^*_\gamma \to G^*_\gamma$ may be normed in such a way that $d\theta = id$.

NOTATION.

$x^*_\gamma(u)$ denotes the image of uX^*_γ (or uZ^*_γ) under the normed isomorphism $\theta : \underline{g}^*_\gamma \to G^*_\gamma$.

So x^*_γ is a k-homomorphism $\underline{G}_a \to G^*_\gamma$, where $\underline{G}_a$ denotes the 1-dimensional additive group, as usual. We have $hx^*_\gamma(u)h^{-1} = x^*_\gamma(h^\gamma u)$ for $h \in T^*$, $u \in K$ (h^γ denotes the image of h under γ).

11.7 Now let γ be a short root. We identify $\underline{g}^*_\gamma$ with its tangent space in 0. Let $\theta : \underline{g}^*_\gamma \to G^*$ be the isomorphism from Proposition 11.5, (ii). Then $h\theta(uX^*_\gamma)h^{-1} = \theta(h^\gamma uX^*_\gamma)$. Differentiating this relation we get $Ad(h)(d\theta)X^*_\gamma = h^\gamma(d\theta)X^*_\gamma$. So $d\theta$ leaves $\underline{g}^*_\gamma$ invariant and θ can be normed in such a way that $d\theta = id$ (note that $d\theta$ is non-zero because θ is an isomorphism).

NOTATION.

The image of uX^*_γ under the normed isomorphism $\theta : \underline{g}^*_\gamma \to G^*$ is denoted $y^*_\gamma(u)$.

So y^*_γ is a morphism $K \to G^*$ satisfying $y^*_\gamma(0) = 1$ and $hy^*_\gamma(uX^*_\gamma)h^{-1} = y^*_\gamma(h^\gamma uX^*_\gamma)$ for $h \in T^*$, $u \in K$. It is not a homomorphism because $(X^*_\alpha)^{[p]} \neq 0$ (see 6.3, Remark 3).

11.8 LEMMA. (cf. [7], Exp. 17, Lemme 1).

Let $\gamma,\gamma_1,\gamma_2,\ldots,\gamma_m \in \Gamma$. Let $f : K^m \to K$ be a morphism, satisfying $h^\gamma f(u_1,\ldots,u_m) = f(h^{\gamma_1}u_1,\ldots,h^{\gamma_m}u_m)$ for $h \in T^*$, $u_1,\ldots,u_m \in K$. Then f is a linear combination of monomials $u_1^{n_1}\ldots u_m^{n_m}$ satisfying $\gamma = n_1\gamma_1 + \ldots + n_m\gamma_m$.

PROOF. Use independence of characters.

11.9 Lemma 11.8 is usually applied in the case that f is the composite of a morphism and a coordinate function. More precisely, if V is an affine variety with coordinates $y_1,\ldots,y_r$ (so $V \subset K^r$), and $\tau : K^m \to V$ is a morphism, then we take $f = y_i \circ \tau$, applying the Lemma r times. Of course this only makes sense if the $y_i \circ \tau$ are nice.

11.10 DEFINITION.

Let Ω be the open cell in G (see 2.1). Then we call $\Omega^* = \phi^{-1}(\Omega)$ the open cell of G^*.

11.11 LEMMA (P1).

(i) Let $\alpha \in \Sigma$. Then $\phi(x_\alpha^*(u))$ (or $\phi(y_\alpha^*(u))$) is equal to $x_\alpha(u)$.

(ii) Let γ be degenerate. Then $\phi(x_\gamma^*(u)) = 1$.

PROOF.

First let $\alpha \in \Sigma$. The inverse image of Ω^* under $\theta : \underline{g}_\alpha^* \to G^*$ is an open T^*-invariant neighbourhood of 0 in $\underline{g}_\alpha^*$. Hence it is $\underline{g}_\alpha^*$ and we have $\phi \circ \theta : \underline{g}_\alpha^* \to \Omega$. Applying Lemma 11.8 it follows from the structure of Ω (see proof of 9.6 or [8], Proposition 1) that $\phi \circ \theta(uX_\alpha^*) = x_\alpha(cu)$, $c \in K$. Differentiating shows that $c = 1$. Part (ii) is proved in the same way.

11.12 The torus T^* acts in a natural way on the direct product of the groups $Z_{G^*}(T^*)$ and $\sum_{\gamma\neq 0} \underline{g}^*_\gamma$, where $Z_{G^*}(T^*)$ denotes the centralizer of T^* in G^*. The action is trivial on the first factor and it is Ad_{G^*} on the second one. We identify the factors with subspaces of the direct product in the natural way.

PROPOSITION (P1). (cf. [7], Exp. 15, Prop. 1).

<u>There is a</u> T^*-<u>equivariant</u> k-<u>isomorphism of varieties</u>

$\theta : Z_{G^*}(T^*) \times \sum_{\gamma\neq 0} \underline{g}^*_\gamma \to \Omega^*$, <u>such that</u>

(i) <u>The restriction of</u> θ <u>to the first factor is the natural embedding</u> $Z_{G^*}(T^*) \to G^*$,

(ii) <u>The restriction to</u> $\underline{g}^*_\gamma$ $(\gamma \neq 0)$ <u>is the normed isomorphism from</u> 11.6 <u>or</u> 11.7,

(iii) <u>There is an order of the non-zero weights of</u> $\underline{g}^*$, <u>say</u> $\beta_1,\dots,\beta_r$, <u>such that</u> $\theta(X_1 + \dots + X_r) = \theta(X_1) \dots \theta(X_r)$ <u>for</u> $X_i \in \underline{g}^*_{\beta_i}$,

(iv) $\theta(x,X) = \theta(x)\theta(X)$ <u>for</u> $(x,X) \in Z_{G^*}(T^*) \times \sum_{\gamma\neq 0} \underline{g}^*_\gamma$.

PROOF.

First we consider R_u. In ([3], 9.12) it is proved that there is a T^*-equivariant isomorphism (over K) $\zeta : \underline{r}_u \to R_u$ and a decomposition of $\underline{r}_u$ into 1-dimensional T^*-stable subspaces L_i, such that, if $L_{(s)}$ denotes $\sum_{i=s}^{m} L_i$, we have

(a) $L_{(1)} = \underline{r}_u$,

(b) For each s, $1 \leq s \leq m$, $\zeta(L_{(s)})$ is a normal subgroup of R_u,

(c) For each s, $1 \leq s \leq m$, the group $\zeta(L_{(s)})/\zeta(L_{(s+1)})$ is T^*-equivariantly isomorphic to L_s.

Now we choose for each s a T^*-equivariant cross section θ_s (over K) of the composite map $\zeta(L_{(s)}) \to \zeta(L_{(s)})/\zeta(L_{(s+1)}) \to L_s$ (see [3], 9.13).

Put $\theta(X_1 + \ldots + X_m) = \theta_1(X_1) \ldots \theta_m(X_m)$ for $X_i \in L_i$. It is clear that θ is an isomorphism of varieties $\underline{r}_u \to R_u$. If γ is a degenerate sum, then it follows from Lemma 11.8 that $\theta^{-1}(x^*_\gamma(u)) = cuZ^*_\gamma$ for some $c \in K$. Hence we may and shall replace the corresponding θ_i by x^*_γ. If $z \in Z_{G^*}(T^*)$, then it follows from the same Lemma that $u \mapsto z\, x^*_\gamma(u)z^{-1}$ is a morphism $K \to R_u$ of the type $u \to x^*_\gamma(cu)$. Hence we may assume that zero weights correspond to the first θ_i. Then we get an isomorphism of varieties from $Z_{R_u}(T^*) \times (\Sigma_{\gamma\ \mathrm{degenerate}}\underline{g}^*_\gamma)$ onto R_u. This isomorphism τ is T^*-equivariant and defined over k. Choose $\beta_1,\ldots,\beta_t$ to be the degenerate sums in the order they occur in the L_i. Choose $\beta_{t+1},\ldots,\beta_r$ to be the roots in ascending order. Then define θ by (i), (ii), (iii), (iv). It has yet to be shown that θ is an isomorphism, as it is clear that θ is T^*-equivariant and defined over k. First we note that $\phi \circ \theta(Z_{G^*}(T^*)) = Z_G(T) = T$.
As T normalizes the subgroups $\{x_\alpha(u) | u \in K\}$, it follows that θ has its image in Ω^* (use Lemma 11.11).
Note that τ is a restriction of θ. The restriction of θ to $Z_{G^*}(T^*) \times (\Sigma_{\gamma\ \mathrm{degenerate}}\underline{g}^*_\gamma)$ is injective because $Z_{G^*}(T^*) \cap \tau(\Sigma_{\gamma\ \mathrm{degenerate}}\underline{g}^*_\gamma) = 1$ (use that τ is T^*-equivariant).
It is an isomorphism because the composite homomorphism $Z_{G^*}(T^*) \to Z_{G^*}(T^*)/Z_{R_u}(T^*) \cong (Z_{G^*}(T^*).R_u)/R_u$ has a rational cross section (see [19], Corollary 1 to Theorem 1 and [1], Proposition 9.4, 6.7). The image of this isomorphism is the connected subgroup $\phi^{-1}(T)$. Note that $\phi^{-1}(T)$ is also connected in the situation of 11.1 (see proof of Lemma 7.4 and use [1], 13.17 Corollary 2, (d)). The result now follows from the structure of Ω (cf. 11.11; reconstruct from $\theta(x,X)$ the components of $(d\phi)X$).

11.13 LEMMA (P1).

Let T^* act on a vector space A such that 0 is contained in the closure of every orbit. Let $\tau : A \to G^*$ be a T^*-equivariant morphism, satisfying $\tau(0) = 1$. Then the image of τ is contained in Ω^*.

PROOF.

$\tau^{-1}(\Omega^*)$ is a T^*-equivariant neighbourhood of 0 in A.

11.14 Let $\tau : A \to G^*$ be given as in the Lemma.
Then we may apply Lemma 11.8 as indicated in 11.9, taking $V = \Omega^*$. We have to choose suitable coordinates on Ω^*. They can be obtained from coordinates on $Z_{G^*}(T^*) \times \sum_{\gamma\neq 0} \underline{g}^*_\gamma$ by the isomorphism θ (see Proposition 11.12). On the factor $Z_{G^*}(T^*)$ we choose some set of coordinates and on the factor $\sum_{\gamma\neq 0} \underline{g}^*_\gamma$ we choose linear coordinates corresponding to the weights. We get results like those in Lemma 11.11, where the same method was applied with Ω instead of Ω^*.

11.15 PROPOSITION (P1).

Let α be a short root.

(i) $(u,v) \mapsto y^*_\alpha(u)x^*_{p\alpha}(v)$ is a k-isomorphism of varieties from K^2 into G^*.

(ii) $y^*_\alpha(a)x^*_{p\alpha}(b)y^*_\alpha(c)x^*_{p\alpha}(d) = y^*_\alpha(a+c)x^*_{p\alpha}(\varepsilon_\alpha f(a,c) + b + d)$, where $\varepsilon_\alpha \in k$ and f is a Witt-cocycle (i.e. $f(a,c) = ac$ if $p = 2$, $f(a,c) = a^2c + ac^2$ if $p = 3$, see [11], p. 197).

(iii) $(X^*_\alpha)^{[p]} = -\varepsilon_\alpha Z^*_{p\alpha}$.

REMARK.

In fact $\varepsilon_\alpha = \pm 1$ in $\underline{g}^*$, as one sees from the proof of 6.2. But this depends on more than (P1) as one sees from example 3 in 11.1 where we have $\varepsilon_\alpha = 0$.

PROOF OF THE PROPOSITION.

(i) The map $(u,v) \mapsto \theta^{-1}(y^*_\alpha(u)x^*_{p\alpha}(v))$ is of the type $(u,v) \mapsto c_1 uX^*_\alpha + c_2 u^p Z^*_{p\alpha} + c_3 vZ^*_{p\alpha}$ (use Lemma 11.8, cf. 11.14). It is clear that $c_1 \neq 0$, $c_3 \neq 0$. Hence it is an isomorphism.

(ii) We argue as in 11.14 and apply Lemma 11.11 (i) and the fact that $x^*_{p\alpha}$ is a homomorphism. As a result we get that the left hand side is equal to $y^*_\alpha(a+c)x^*_{p\alpha}(h(a,c)+b+d)$, where h is a homogeneous polynomial of degree p. It follows that h(a,c) is a 2-cocycle of $\underline{G}_a$ in $\underline{G}_a$ (with trivial action). Hence we can apply ([11], II §3 n° 4.6) to see that h is spanned by polynomials of the form fP^r, $(XY^{p^r})^{p^n}$, $X^n+Y^n - (X+Y)^n$, where $n,r \geqslant 0$, f is a Witt-cocycle. But f is the only one with degree p.

(iii) As p = 2 or 3, we have $(y^*_\alpha(u))^p = x^*_{p\alpha}(-\varepsilon_\alpha u^p)$. The group generated by the elements $y^*_\alpha(u)$, $x^*_{p\alpha}(u)$ is solvable. So it can be realized in trigonalized form. In that form (iii) is an easy consequence of the relation $(y^*_\alpha(u))^p = x^*_{p\alpha}(-\varepsilon_\alpha u^p)$.

11.16 LEMMA (P1).

<u>Let</u> $\gamma \neq 0$ <u>be a weight of</u> $\underline{g}^*$, ψ_γ <u>a</u> T^*-<u>equivariant morphism from</u> $\underline{g}^*_\gamma$ <u>into</u> G^*, <u>mapping</u> 0 <u>to</u> 1.

(i) $d\psi_\gamma$ <u>maps</u> $\underline{g}^*_\gamma$ <u>into itself</u>.

(ii) <u>If</u> $d\psi_\gamma = c$ id, $c \in K$, <u>and</u> γ <u>is not a short root, then</u> $\psi_\gamma(X) = \theta(cX)$ <u>for all</u> $X \in \underline{g}^*_\gamma$. <u>Here</u> θ <u>is the isomorphism that defines</u> x^*_γ (<u>see</u> 11.6).

(iii) <u>If</u> $d\psi_\gamma = c_1$id, $c_1 \in K$, <u>and</u> γ <u>is a short root, then there is</u> $c_2 \in K$ <u>such that</u> $\psi_\gamma(uX^*_\gamma) = y^*_\gamma(c_1 u)x^*_{p\gamma}(c_2 u^p)$. <u>If</u> ψ_γ <u>is defined over</u> k, <u>then</u> $c_1, c_2 \in k$.

PROOF.

(i) See 11.7.

(ii) Note that θ in Proposition 11.12 extends $\theta : \underline{g}^*_\gamma \to G^*$. The result is obtained by the argument in 11.14.

(iii) Use the same method.

11.17 DEFINITION.

Let α be a short root, $c_\alpha \in k$. Then we put $x^*_\alpha(u) = y^*_\alpha(u)x^*_{p\alpha}(c_\alpha u^p)$. We say that x^*_α is obtained from y^*_α by the norming constant c_α. From now on a set of norming constants is supposed to be given.

REMARK.

It follows from Lemma 11.16 (iii) that the norming constants represent the freedom of choice in the definition of y^*_α (see Proposition 11.5 (ii) and 11.7). Hence results like Proposition 11.15 are also valid when y^*_α is replaced by x^*_α. We will use this frequently.

11.18 PROPOSITION (P1). (cf. [22], Lemma 15).

Let α,β be independent weights of $\underline{g}^*$.

(i) $(x^*_\alpha(u),x^*_\beta(v)) = \prod_{i>0,j>0} x^*_{i\alpha+j\beta}(c_{ij\alpha\beta}u^iv^j)$, where the product is taken in any order and the $c_{ij\alpha\beta}$ are elements of k (depending on the order).

(ii) We fix the order of the product in (i). If i,j are not both divisible by p, then $c_{ij\alpha\beta}$ can be determined from the action of the elements $x^*_\gamma(t)$ $(t \in K,\ \gamma \neq 0)$ on the weight spaces $\underline{g}^*_\delta$ with δ linearly independent from γ (see 7.2 for the action).

REMARKS.

1) If $c_{ij\alpha\beta} \neq 0$ and $i\alpha+j\beta$ is a short root, then $c_{pi,pj,\alpha,\beta}$ depends on the norming constants. So the condition in (ii) is essential.

2) $c_{11\alpha\beta}$ corresponds to a commutator in the Lie algebra (cf. [22], Lemma 15).

3) If $x^*_\alpha(u) \in R_u$ or $x^*_\beta(v) \in R_u$, then we only have to use weights $i\alpha+j\beta$ that are degenerate.

PROOF OF THE PROPOSITION.

(i) First take the same order of the weights as in Proposition 11.12, (iii). Then the result follows as above (see 11.14). For an arbitrary order we reason by induction on the number of weights $i\alpha+j\beta$ ($i > 0$, $j > 0$) that occur in $\underline{g}^*$. By induction hypothesis every product $\Pi x^*_{i\alpha+j\beta}(u_{ij})$ can be reordered using (i) for commutators $(x^*_{i\alpha+j\beta}(u_{ij}), x^*_{r\alpha+s\beta}(u_{rs}))$ (cf. [22], p.24-26).

(ii) Let G^* be realized as a linear algebraic group, $G^* \subset GL_n$. Then we can multiply matrices in G^* with matrices in $\underline{g}^*$, and we can differentiate morphisms $K^n \to G^*$ in the same way as we differentiate polynomials. (In fact they are polynomials with matrices as coefficients.)

If γ is a short root, then it follows from

$x^*_\gamma(u+v) = x^*_\gamma(u)\, x^*_{p\gamma}(-\varepsilon_\gamma f(u,v))\, x^*_\gamma(v)$ that

$(\frac{d}{du} x^*_\gamma(u+v))_{u=0} = X^*_\gamma x^*_\gamma(v) - \varepsilon_\gamma v^{p-1} Z^*_{p\gamma} x^*_\gamma(v)$.

So $v\frac{d}{dv}(x^*_\gamma(v)) = (vX^*_\gamma - \varepsilon_\gamma v^p Z^*_{p\gamma}) x^*_\gamma(v)$.

For long roots and for degenerate sums one gets analogous formulas. Now we note that $xX = (Ad_{G^*}(x)X)x$ for $x \in G^*$, $X \in \underline{g}^*$. Hence elements of $\underline{g}^*$ can be "transported to the left" and we can apply the same method as Steinberg used in ([22], proof of Lemma 11.18). Applying $u\frac{d}{du}$ to both sides of (i) we get relations that enable us to determine inductively all $c_{ij\alpha\beta}$ with i prime to p (induction on $i+j$). Applying $v\frac{d}{dv}$ to both sides we get the same kind of relations with j prime to p.

11.19 DEFINITION.

Let α,β be independent weights. Put

$$G^*_{(\alpha,\beta)} = \{ \prod_{\substack{i\geq 0, j\geq 0\\ i+j>0}} x^*_{i\alpha+j\beta}(v_{ij}) \mid v_{ij} \in K\},$$ where the product is taken in some fixed order, skipping $i\alpha+j\beta$ if it is not a weight of $\underline{g}^*$.

COROLLARY (P1).

(i) $G^*_{(\alpha,\beta)}$ <u>is a</u> k-<u>subgroup of</u> G^*.

(ii) <u>There is a bijective correspondence between the elements of</u> $G^*_{(\alpha,\beta)}$ <u>and their parameters</u> v_{ij}.

(iii) <u>This correspondence is a</u> k-<u>isomorphism</u> $K^m \to G^*_{(\alpha,\beta)}$, <u>of algebraic varieties, where</u> $m = \dim G^*_{(\alpha,\beta)}$.

PROOF.

This Corollary may be proved in the same way as Proposition 11.15. Part (i) also follows from Proposition 11.18 (i), using Proposition 11.15 (ii) and 11.5, 11.6. Parts (ii), (iii) follow from Propositions 11.12, 11.18 (i) (cf. [22], p. 24-26).

REMARK.

It follows from part (i) of the Corollary that $G^*_{(\alpha,\beta)}$ does not depend on the order that is used in its definition.

11.20 Given some expression $x^*_{\gamma_1}(u_1) \ldots x^*_{\gamma_n}(u_n)$ it is often possible to reorder the factors such that the weights occur in some prescribed order. That is: $x^*_{\gamma_1}(u_1) \ldots x^*_{\gamma_n}(u_n) = x^*_{\delta_1}(v_1) \ldots x^*_{\delta_r}(v_r)$, where the δ_i are ordered in the prescribed way. Applying Proposition 11.18 several times one may try to express the arguments v_i in terms of the u_j and the constants of type $c_{ij\alpha\beta}$. It can be done for instance in the case that all factors are contained in a subgroup of type

$G^*_{(\alpha,\beta)}$. We call the technique "Reordering the product".

11.21 THEOREM.

<u>Let</u> G <u>not be of type</u> B_3, $\phi : G^* \to G$ <u>as above (see</u> 11.4). <u>Then</u> R_u <u>is commutative</u>.

PROOF.

R_u is solvable. If $\underline{r}_u$ is an irreducible G-module, then (R_u,R_u) has trivial Lie algebra and is connected, so $(R_u,R_u) = \{1\}$. So we are done in the case of type F_4 (see Proposition 5.2). Hence we may suppose that Σ is not of type F_4. Then $\dim(\underline{r}_u)_0 \leq 1$ (see Proposition 5.2 again). Let $Z(T^*)$ denote the centralizer of T^* in G^*. The group $Z(T^*) \cap R_u$ is a connected group of dimension ≤ 1 (see [1], (9.4)), hence it is abelian. (see [1], (10.9)). If $z \in Z(T^*) \cap R_u$, and γ is a degenerate sum, then $uZ^*_\gamma \mapsto (z, x^*_\gamma(u))$ satisfies the conditions of Lemma 11.16. Its derivative $Ad(z) - id = \widehat{Ad}(\phi z) - id$ is trivial (cf. Lemma 11.2), so

(1) $Z(T^*) \cap R_u$ is central in R_u.

Next we consider two independent degenerate sums γ,δ. There is no degenerate sum in $p^2\Gamma$ (see Lemma 2.6 (i)), so we can apply Proposition 11.18 (ii) to see that the constants $c_{ij\gamma\delta}$ are zero. (They are zero in one solution of $d\phi = \pi$ because of Theorem 10.1, so they must be zero in any solution). So

(2) $x^*_\gamma(u)$ commutes with $x^*_\delta(v)$ if $\gamma+\delta \neq 0$.

Now we have to consider the case $\gamma+\delta = 0$.

EXAMPLE.

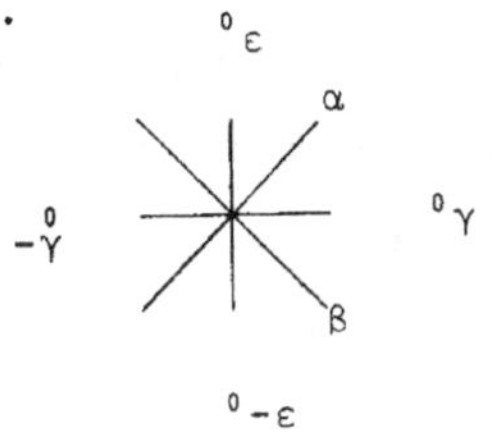

α,β are long roots, $p = 2$, γ,ε are degenerate sums, see figure.

We apply Proposition 11.18 with explicit constants $c_{ij\alpha\beta}$. These constants are obtained from known solutions of $d\phi = \pi$ (see section 10 and use Proposition 11.18 (ii)) or as indicated in the proof of 11.18 (ii). We get

$Int(x^*_\gamma(u))\ x^*_{-\gamma}(v) = Int((x^*_\alpha(u),\ x^*_\beta(1)))x^*_{-\gamma}(v) =$

$Int(x^*_\alpha(u)\ x^*_\beta(1)\ x^*_\alpha(u))x^*_{-\gamma}(v)\ x^*_{-\varepsilon}(v) =$

$Int(x^*_\alpha(u)\ x^*_\beta(1))\ x^*_\varepsilon(u^2v)\ x^*_{-\gamma}(v)\ x^*_{-\varepsilon}(v)\ x^*_\gamma(u^2v) =$

$Int(x^*_\alpha(u))\ x^*_\varepsilon(u^2v)\ x^*_\gamma(u^2v)\ x^*_{-\gamma}(v)\ x^*_{-\varepsilon}(v)\ x^*_{-\varepsilon}(v)\ x^*_\gamma(u^2v) =$

$x^*_\varepsilon(u^2v)\ x^*_\gamma(u^2v)\ x^*_{-\gamma}(v)\ x^*_\varepsilon(u^2v)\ x^*_\gamma(u^2v) =$

$Int(x^*_\gamma(u^2v))\ x^*_{-\gamma}(v)$, where $Int(x)y = xyx^{-1}$ as usual.

Put $f(u,v) = (x^*_\gamma(u), x^*_{-\gamma}(v))$. Then f is a morphism, satisfying $f(u,v) = f(u^2v,v)$ and $f(0,v) = 1$. It is easy to see that f is constant (use coordinate functions).

If $p = 3$ then the same method can be applied, without knowledge of the signs of the $c_{ij\alpha\beta}$. If G is of type B_3, then the trick fails however, because i,j are both even in some relevant $c_{ij\alpha\beta}$ $(p = 2)$. Then we can't apply Proposition 11.18.It seems that this case is difficult because degenerate sums of two distinct lengths occur. If G is of type G_2, $p = 2$, then there are also some relevant constants of type $c_{2i,2j,\alpha,\beta}$. We shall handle this case separately in 11.24, 11.25. It is easily seen from 2.8 Table 1 that there is no other case then those mentioned above. So now we exclude types G_2 and B_3 in characteristic 2. Then it follows from (1), (2) and the relation $f = 1$ in the example that

(3) $x^*_\gamma(u)$ is central in R_u.

The Theorem follows from (1), (3).

The proof for case G_2, $p = 2$, will be given in 11.25.

11.22 LEMMA (P1).

(R_u,R_u) <u>is contained in</u> $Z(T^*) \cap R_u$.

PROOF.

As in the proof of 11.21 we see that $x^*_\gamma(u)$ commutes with $Z(T^*) \cap R_u$ and with $x^*_\delta(v)$, where γ,δ are degenerate, $\gamma+\delta \neq 0$. So (R_u,R_u) is generated by $(R_u,R_u) \cap Z(T^*)$ and by the commutators $(x^*_\gamma(u),x^*_{-\gamma}(v))$, γ degenerate.

We use the isomorphism $Z_{R_u}(T^*) \times \sum_{\gamma \text{ degenerate}} \mathfrak{g}^*_\gamma \to R_u$ (see 11.18) and Lemma 11.8 to see that the commutators $(x^*_\gamma(u),x^*_{-\gamma}(v))$ are contained in $Z_{R_u}(T^*)G^*_\gamma G^*_{-\gamma}$ or $Z_{R_u}(T^*)G^*_{-\gamma}G^*_\gamma$ (notations as in 11.5). Suppose they are not contained in $Z_{R_u}(T^*)$. Then (R_u,R_u) contains one of the groups $G^*_\gamma,G^*_{-\gamma}$(see Lemma 11.16, [1] Proposition 9.4, [3] Theorem 9.16, cf. proof of 11.5). But R_u is nilpotent (see [1], Corollary 10.5), whence a contradiction.

11.23 LEMMA (P1).

<u>Let</u> α <u>be a short root</u>.

$(x^*_\alpha(u),x^*_{-p\alpha}(v)) = x^*_{p\alpha}(\pm vu^{2p})\tau^\alpha(vu^p)$, <u>where</u> τ^α <u>is a morphism</u> $K \to Z(T^*) \cap R_u$

PROOF.

The map $f : (u,v) \mapsto \mathrm{Int}(x^*_\alpha(u))x^*_{-p\alpha}(v)$ has its image in R_u.

Applicating Lemma 11.8 as in 11.22 we see

(1) $f(u,v) = x^*_{-p\alpha}(vf_1(vu^p))\tau^\alpha(vu^p)x^*_{p\alpha}(vu^{2p}f_2(vu^p))$, where $\tau^\alpha(vu^p) \in Z(T^*) \cap R_u$.

If R_u is not commutative, then we replace G^* by $G^*/(R_u,R_u)$.

This makes sense because of Lemma 11.22.

The action Int of G^* on $R'_u = R_u/(R_u,R_u)$ factors through G.

This yields an action ρ of G on R'_u.

Now a standard argument shows

(2) $\rho(w_{-\alpha}(t))(R'_u)_\gamma \subset (R'_u)_{\gamma-\langle\gamma,\alpha\rangle\alpha}$ (see [2], 3.3, Remark 1).

Put $g(u,v) = \mathrm{Int}(x^*_{-\alpha}(u))x^*_{p\alpha}(v)$. Then

(3) $g(u,v) = x^*_{-p\alpha}(u^{2p}vg_1(vu^p))\sigma(vu^p)x^*_{p\alpha}(vg_2(vu^p))$, where σ is a morphism $K \to Z(T^*) \cap R_u$ (cf. (1)).

In the same way

(4) $\mathrm{Int}(x^*_{-\alpha}(u))\tau^\alpha(v) = x^*_{-p\alpha}(u^p h(v))\tau'(v)$, where τ' is a morphism $K \to Z(T^*) \cap R_u$.

Substituting $u = 0$ one sees $\tau' = \tau^\alpha$. We get modulo (R_u,R_u):

$\rho(w_{-\alpha}(t))x^*_{-p\alpha}(u) = \mathrm{Int}(x^*_{-\alpha}(t)x^*_\alpha(-t^{-1})x^*_{-\alpha}(t))x^*_{-p\alpha}(u) = x^*_{p\alpha}(l(t,u))r$,

where $l(t,u) = ut^{-2p}f_2(-t^{-p}u)g_2(ut^{-p}f_2(-t^{-p}u))$, r corresponds to other weights then $p\alpha$.

From (2) it follows that $u \mapsto \rho(w_{-\alpha}(t))x^*_{-p\alpha}(u)$ is an invertible homomorphism $K \to (R'_u)_{p\alpha}$ (see 11.6). So l is linear in u and $f_2(x)g_2(-xf_2(x))$ is a non-zero constant ($x = -t^{-p}u$). Then f_2 is a non-zero constant and g_2 is a non-zero constant. Similarly f_1 and g_1 are constant. Their values are obtained by differentiating f and g with respect to v.

11.24 LEMMA (P1).

Let G be of type G_2, $p = 2$.

If δ is degenerate, ζ is a root, then $c_{2,2,\delta,\zeta} = 0$.

PROOF.

We use the same notations for the roots as in 10.15.

$\mathrm{Int}(x^*_{\beta-\alpha}(t))x^*_{2\alpha}(u) = \mathrm{Int}((x^*_\beta(t),x^*_{-\alpha}(1)))x^*_{2\alpha}(u)$. Write the right hand side as a product and reorder it, using Lemma 11.23 (see 11.20). The result has no component in $G^*_{-2\gamma}$. So $c_{2,2,+2\alpha,\beta-\alpha} = 0$. Other cases are of the same type.

11.25 We finish the proof of Theorem 11.21.

From Lemma 11.24 it follows that we can handle G_2 in the same way as we handled other cases. Note that the same would be true in the case B_3, $p = 2$, if we could prove $c_{2,2,-\varepsilon_1-\varepsilon_2,\varepsilon_1+\varepsilon_2+\varepsilon_3}$ to be zero.

11.26 Let R_u be commutative. Then the action Int of G^* on R_u factors through G.

NOTATION.

The resulting action of G on R_u is denoted $\hat{\text{I}}\text{nt}$.

There is also the action $\hat{\text{A}}\text{d}$ of G on $\underline{r}_u$, satisfying $\text{Ad}_{G^*}(x) = \hat{\text{A}}\text{d}(\phi x)$ for $x \in G^*$. The derivative of $y \mapsto \hat{\text{I}}\text{nt}(\phi x)(y)$ is $\hat{\text{A}}\text{d}(\phi x)$.

11.27 THEOREM. (cf. 11.1, (P2)).

Let R_u be commutative. Then there is a G^*-equivariant separable k-homomorphism τ from $\underline{r}_u$ onto R_u. Its finite kernel spans a G^*-invariant subspace of dimension $\leqslant 1$.

PROOF.

We define τ in the following way.

(1) The restriction of τ to $\Sigma_{\gamma \text{ degenerate}} \underline{g}^*_\gamma$ is equal to the restriction of θ (see Proposition 11.12). If there is a short root choose one, say α. Define Z^α by the relation

(2) $\hat{\text{A}}\text{d}(x_\alpha(t))Z^*_{-p\alpha} = Z^*_{-p\alpha} + t^p Z^\alpha \pm t^{2p} Z^*_{p\alpha}$.

Then put

(3) $\tau(uZ^\alpha) = \tau^\alpha(u)$ (see Lemma 11.23).

If G is of type F_4, choose a short root β, such that the angle between α and β is $\frac{2\pi}{3}$.

Then we put

(4) $\tau(uZ^{\beta}) = \tau^{\beta}(u)$, where Z^{β}, τ^{β} are the analogues of Z^{α},τ^{α}. From (1), (2), (3), (4) we get a consistent definition of the homomorphism τ (see Corollary 3.14 and Proposition 5.2). It is obvious that τ is a k-homomorphism from $\underline{r}_u$ into R_u. Next we show that τ is G^*-equivariant. Equivalently, we show that τ is G-equivariant. As generators of G we take the $x_\delta(t)$ with δ long together with $x_\alpha(t)$, $x_\beta(t)$ (if existent) with α,β as above. First consider $\hat{\text{I}}\text{nt}\,(x_\delta(t))$. Its action on $Z(T^*) \cap R_u$ is trivial because of Lemma 11.8 (cf. 11.21 proof of (1)). If γ is degenerate, then $\hat{\text{I}}\text{nt}(x_\delta(t))x_\gamma(u)$ can usually be determined from Proposition 11.18, Lemma 11.24. We claim that the only exception is type B_3, $p = 2$. To prove the claim, let Σ not be of type B_3 or G_2 or let $p \neq 2$. Let $pi\delta + pj\gamma$ be degenerate $(i > 0, j > 0)$. Then $(\gamma,\gamma) = (pi\delta + pj\gamma, pi\delta + pj\gamma) = p(\delta,\delta)$ (see the classification of degenerate sums in section 2). So
$p(\delta,\delta) = 2p^2ij(\gamma,\delta) + p^2i^2(\delta,\delta) + p^2j^2(\gamma,\gamma) =$
$p(\delta,\delta)\{pij\langle\gamma,\delta\rangle + pi^2 + p^2j^2\}$ which is nonsense.

So we may assume that G is of type B_3 and that $\gamma = \varepsilon_1+\varepsilon_2+\varepsilon_3$, $\delta = -\varepsilon_2-\varepsilon_3$, $i = j = 1$. Then

$$\hat{\text{I}}\text{nt}(x_{-\varepsilon_1}(t))\hat{\text{I}}\text{nt}(x_{-\varepsilon_2-\varepsilon_3}(u))\, x^*_{\varepsilon_1+\varepsilon_2+\varepsilon_3}(v) =$$

$$x^*_{\varepsilon_1+\varepsilon_2+\varepsilon_3}(v)\; x^*_{-\varepsilon_1+\varepsilon_2+\varepsilon_3}(t^2v)\; x^*_{2\varepsilon_1}(c_{2,2,\gamma,\delta}u^2v^2) \times$$

$$\tau^{-\varepsilon_1}(c_{2,2,\gamma,\delta}t^2u^2v^2)x^*_{-2\varepsilon_1}(c_{2,2,\gamma,\delta}t^4u^2v^2)x^*_{\varepsilon_1-\varepsilon_2-\varepsilon_3}(u^2v) \times$$

$$x^*_{-\varepsilon_1-\varepsilon_2-\varepsilon_3}(t^2u^2v).$$

But

$$x_{-\varepsilon_1}(t)x_{-\varepsilon_2-\varepsilon_3}(u) = x_{-\varepsilon_2-\varepsilon_3}(u)x_{-\varepsilon_1}(t) \text{ and}$$

$$\hat{\text{I}}\text{nt}(x_{-\varepsilon_2-\varepsilon_3}(u))\hat{\text{I}}\text{nt}(x_{-\varepsilon_1}(t))x^*_{\varepsilon_1+\varepsilon_2+\varepsilon_2}(v) =$$

some expression that lacks the component with weight zero (use that R_u is commutative). So $\tau^{-\varepsilon_1}(c_{2,2,\gamma,\delta}t^2u^2v^2) = 1$. But $\tau^{-\varepsilon_1}$ is non-trivial (take derivatives), so $c_{2,2,\gamma,\delta} = 0$. So far about the action of $x_\delta(t)$.

Next consider the action of $x_\alpha(t)$. It is seen from $\hat{\mathrm{I}}\mathrm{nt}(x_\alpha(t+u))x^*_{-p\alpha}(v) = \hat{\mathrm{I}}\mathrm{nt}(x_\alpha(t))\hat{\mathrm{I}}\mathrm{nt}(x_\alpha(u))x^*_{-p\alpha}(v)$ that $x_\alpha(t)$ acts in the right way on $\tau^\alpha(u^pv)$. In the same way it follows (if β exists) from $(\hat{\mathrm{I}}\mathrm{nt}(x_\beta(t)),\hat{\mathrm{I}}\mathrm{nt}(x_\alpha(u)))x^*_{-2\beta}(v) = \hat{\mathrm{I}}\mathrm{nt}(x_{\alpha+\beta}(tu))x^*_{-2\beta}(v)$ that $x_\alpha(u)$ acts in the right way on $\tau^\beta(t^pv)$. The action of $x_\alpha(u)$ on $x^*_{\pm p\alpha}(v)$ poses no problem. We claim that $\hat{\mathrm{I}}\mathrm{nt}(x_\alpha(u))x^*_\gamma(v)$ can be determined from $\hat{\mathrm{A}}\mathrm{d}$ if γ is a degenerate sum distinct from $\pm p\alpha$. So we claim that no $c_{pi,pj,\alpha,\gamma}$ occurs (see Proposition 11.18 (ii)). Suppose it did. Then there are $i > 0$, $j > 0$ such that $pi\alpha + pj\gamma$ is degenerate. This doesn't occur in type B_3. If Σ is not of type B_3, then $(pi\alpha + pj\gamma,\ pi\alpha + pj\gamma) = (\gamma,\gamma) = p^2(\alpha,\alpha)$ (see 2.9 (iii), Lemma 2.11). It follows that $i^2 + ij < \gamma,\alpha > + p^2j^2 = 1$, while $|< \gamma,\alpha >| < p < \alpha,\alpha > = 2p$. And $< \gamma,\alpha > \in p\mathbb{Z}$, so

$1 = i^2 + ij < \gamma,\alpha > + p^2j^2 \geq i^2 - pij + p^2j^2 \geq pij \geq p$.

This is a contradiction.

Summing up, we have seen that $x_\delta(t)$, $x_\alpha(t)$ act in the right way. For reasons of symmetry $x_\beta(t)$ does too. It follows that τ is G-equivariant. Separability follows from the fact that $\mathrm{Im}(d\tau)$ contains generators of $\underline{r}_u$ (see Proposition 5.2). The kernel of τ is a zero-dimensional algebraic group, fixed by G. The Theorem then follows from Proposition 5.2.

11.28 In case B_3 the proof uses the fact that $Z(T^*) \cap R_u$ is non-trivial, in order to get rid of $c_{2,2,-\varepsilon_1-\varepsilon_2,\varepsilon_1+\varepsilon_2+\varepsilon_3}$. So we can't apply the same proof to $G^*/(R_u,R_u)$ in the case that R_u is not commutative. Accounting for that, we get as a corollary to the proof:

11.29 COROLLARY (cf. 11.1, (P2)).

Let $\phi : H \to G$ be given as in 11.1, such that (P1) holds (see 11.1). Let $R_u(H)$ be commutative. If p=2 and G is of type B_3 assume that one of the two orbits of degenerate sums doesn't occur in the weights of $\underline{h}$. Then there is an H-equivariant k-homomorphism $\tau : \underline{r}_u \to R_u(H)$ satisfying $d\tau = \mu$ (see (P1) for μ).

11.30 THEOREM.

Let H be a connected (linear) algebraic group with perfect Lie algebra (i.e. $\underline{h} = [\underline{h},\underline{h}]$). Assume that $p \neq 2$ or that H has no quotient of type B_3. Let the Lie algebra $\underline{r}$ of $R_u(H)$ be central in $\underline{h}$. Then there is an H-equivariant separable homomorphism τ from an H-module M onto $R_u(H)$. If H is defined over k and H has a k-split maximal torus, then τ may be taken to be defined over k.

REMARK.

We may assume that ker τ consists of invariants, because otherwise ker τ contains an H-submodule of M (apply [3], Theorem 9.16 to the semi-direct product ⌜M,H⌝).

PROOF.

Put $G = H/R_u(H)$. (So G is not necessarily the same as above). Then $\underline{g}$ is perfect, because $\underline{h}$ is perfect. So G is semi-simple (see [1], 14.2) and $\underline{g}$ is isomorphic to the Lie algebra of the simply connected group G_1 that covers G (see proof of Lemma 7.1). Then $\underline{g} = \bigoplus_i \underline{g}_i$ where $\underline{g}_i$ denotes the Lie algebra of an almost simple factor G_i of G. We have $\underline{g}^* = \bigoplus_i \underline{g}_i^*$. There is a surjection of Lie algebras $\mu : \underline{g}^* \to \underline{h}$ induced by $\rho : \underline{h} \to \underline{g}$. From the central trick it follows that μ is H-equivariant. We may assume that H is defined over k and that H contains a k-split maximal torus T^*. (Otherwise change k.) Let G_i be a factor of G. For simplicity of notations we assume G_i to be isomorphic to the corresponding subgroup of G. We identify G_i with that subgroup.

Let $\phi : H \to G$ be the canonical homomorphism. The torus $\phi(T^*) = T$ is isomorphic to T^* (ker ϕ is unipotent and ϕ is separable). The subtorus $T_i = T \cap G_i$ corresponds to a subtorus T_i^* of T^* such that $\phi(T_i^*) = T_i$. We may assume that T_i is a maximal torus in G_i (see [1], proof of Theorem 14.10 (3)). Consider the homomorphism $\phi_i : H \to G_i$ and the tori T_i^*, T_i. The situation is that of 11.1 with (P1) because T_i^* (or T_i) acts trivially on $\mu(\underline{g}_j^*)$ for $i \neq j$. Hence we have morphisms $x^*_{\alpha,i} : K \to H$ corresponding to roots in G_i (see Proposition 11.5, Definitions 11.6, 11.7, 11.17). Their images generate a subgroup H_i of H. We claim that H_i commutes with H_j for $i \neq j$. This is proved as follows.

$Z_H(T_i^*)$ contains T_j^* and its Lie algebra contains $\mu(\underline{g}_j^*)$ (see [1], Proposition 9.4). So the $x^*_{\beta,j}$ have their images in $Z_H(T_i^*)$ (see Lemma 11.16 and [3], Theorem 9.16).It follows from Lemma 11.16 (cf. proof of 11.21 or 11.22) that $x^*_{\beta,j}(v)$ commutes with $x^*_{\alpha,i}(u)$. (Apply the Lemma twice).

Now we want to prove that $R_u(H)$ is commutative. In view of the above we may restrict ourselves to the radical of $H^i = H/\prod_{j\neq i} H_j$. But then the situation is just the same as in 11.21, except that H^i may be smaller then G^*, which causes no problem.

We may apply Corollary 11.29 to see that there is a separable H-equivariant k-homomorphism τ' from $\oplus_i \underline{r}_{u,i}$ onto $R_u(H)$, where $\underline{r}_{u,i}$ is "the $\underline{r}_u$ of G_i" (see 11.1).

§12. Representatives in G^* of the Weyl group.

In this section we lift representatives of the Weyl group to elements of G^* normalizing the maximal torus T^*. The main goal is to get the analogue of relation (C) from Steinbergs set of defining relations for G (see [23]).

12.1. We return to the notations of 11.4, using labels as described in 11.3.

12.2. DEFINITIONS.

For $\alpha \in \Sigma$ we put $w^*_\alpha(t) = x^*_\alpha(t)\, x^*_{-\alpha}(-t^{-1})\, x^*_\alpha(t)$ and

$$h^*_\alpha(t) = w^*_\alpha(t)(w^*_\alpha(1))^{-1} \quad (t \in K^\times)$$

(see 2.1).

The group generated by the elements $x^*_{\pm\alpha}(u)$ is denoted $G^{*\alpha}$.

12.3. The image G^α of $G^{*\alpha}$ in G is of type SL_2 (see [2], 3.3(2)). The Lie algebra of $G^{*\alpha}$ has only weights $n\alpha$ ($n \in \mathbb{Z}$), because $G^{*\alpha}$ centralizes $\ker(\alpha : T^* \to K)$. First let α be long. Then $Z(T^*) \cap R_u$ commutes with the elements $x^*_{\pm\alpha}(u)$ (see 11.21 proof of (1)), so $G^{*\alpha} \to G^\alpha$ is a central extension (cf. proof of 11.5(i)). We can apply ([1], 10.9) and Theorem 9.6 to see that there is an inverse homomorphism s. From the central trick for groups it follows that $s(x_\alpha(u)) = x^*_\alpha(u)$ (see proof of 9.6 and use the central extension $T^* \,.\, G^{*\alpha} \to T \,.\, G^\alpha$). Let $h \in T^*$ such that $\phi(h) \in G^\alpha$. Then $h^{-1} \,.\, s(\phi(h))$ is unipotent and commutes with h (consider the same extension). So it is the unipotent part of $s(\phi(h))$ which is zero. Hence $h^*_\alpha(t) = s(\phi(h^*_\alpha(t))) = s(\phi(h)) = h$ for some $h \in T^*$.

12.4. PROPOSTION (P3).

<u>Let</u> α <u>be a long root</u>.

(i) $h^*_\alpha(t) \in T^*$,

(ii) $w^*_\alpha(t)$ <u>normalizes</u> T^*,

(iii) The group $G^{*\alpha}$ is isomorphic to SL_2.

PROOF.

Part (i) and (iii) have been proved above. Part (ii) is easy because $(h,w_\alpha^*(t)) = h_\alpha^*(h^\alpha t)\ h_\alpha^*(t)^{-1}$ for $h \in T^*$.

12.5 PROPOSITION. (P3)

Let α be a short root. For each value (in k) of the norming constant c_α there is a value of $c_{-\alpha}$ (in k) such that

(i) $h_\alpha^*(t) \in T^*$,

(ii) $w_\alpha^*(t)$ normalizes T^*,

(iii) $\mathrm{Int}\ (w_\alpha^*(t))\ x_{-\alpha}^*(-t^{-1}) = x_\alpha^*(t)$.

REMARK. Property (P3) is sufficient, but we need not exclude type B_3.

PROOF.

If (ii) holds, then the usual argument shows that $\mathrm{Int}\ (w_\alpha^*(t))\ x_\beta^*(u) \in G^*_{\beta-<\beta,\alpha>\alpha}$ for long roots β (see [2], (3.3) Remark 1). We want to use the reverse of this implication. Hence we first consider $\mathrm{Int}\ (w_\alpha^*(t))\ x_\beta^*(u)$. Evaluating this expression by "reordering the product" (see 11.20) one has to check whether all factors cancel out whose weights are not $\beta - <\beta,\alpha>\alpha$. For those factors which are linear in u the cancellation follows from the corresponding fact in the Lie algebra, where $w_\alpha^*(t)$ acts in the same way as $w_\alpha(t)$. For the factors corresponding to roots it follows from the corresponding fact in G. So we look at the case that $\pm\ i\alpha + j\beta$ is degenerate, $i > 0$, $j > 1$. Checking the 2 dimensional root systems and using Proposition 2.12 it is seen that there are two possibilities

(a) $\pm\alpha$, β are simple roots in type G_2.

(b) $\pm\alpha$, β are simple roots in a subsystem of type B_2.

In case (a) we argue as follows.

Fix c_α. If one changes $c_{-\alpha}$ by an amount d, then $w^*_\alpha(t)$ is multiplied on the left by $x^*_{p\alpha}(\pm dt^p)\ \tau^\alpha(d)x^*_{-p\alpha}(dt^{-p})$ (see Lemma 11.23). Hence we can choose $c_{-\alpha}$ in such a way that $\mathrm{Int}\ (w^*_\alpha(t))\ x^*_\beta(u) = x^*_{\beta\pm 3\alpha}(\dots)\ x^*_{p(\beta\pm\alpha)}(\dots)$, without a component $x^*_{p(\beta\pm 2\alpha)}(\dots)$. Say

$\mathrm{Int}\ (w^*_\alpha(t))\ x^*_\beta(u) = x^*_{\beta+3\alpha}(t^3u)\ x^*_{p(\beta+\alpha)}(F_1t^pu^p)$, where $F_1 \in K$.

Say furthermore

$$\mathrm{Int}\ (w^*_\alpha(t))\ x^*_{-\beta}(-u^{-1}) = x^*_{-\beta-3\alpha}(\pm t^{-3}u^{-1})\ x^*_{p(-\beta-\alpha)}(F_2t^{-p}u^{-p})\ x^*_{p(-\beta-2\alpha)}(F_3t^{-2p}u^{-p}).$$

Then

(1) $$\mathrm{Int}\ (w^*_\alpha(t))\ w^*_\beta(u) = x^*_{p(\beta+\alpha)}(2F_1t^pu^p)\ x^*_{p\alpha}(\pm F_3t^p)\ x^*_{p(-\beta-\alpha)}(F_2t^{-p}u^{-p})x^*_{p(-\beta-2\alpha)}(F_3t^{-2p}u^{-p})w^*_{\beta+3\alpha}(\pm t^3u).$$

(see [22], Lemma 19).

Now let both sides of (1) act on $x^*_{\beta+3\alpha}(t^3u)\ x^*_{p(\beta+\alpha)}(F_1t^pu^p)$.

That gives

$$x^*_{-\beta-3\alpha}(\pm t^{-3}u^{-1})x^*_{p(-\beta-\alpha)}(F_2t^{-p}u^{-p})x^*_{p(-\beta-2\alpha)}(F_3t^{-2p}u^{-p}) =$$

$$x^*_{-\beta-3\alpha}(-t^{-3}u^{-1})\ x^*_{p(\beta+\alpha)}(F_1t^pu^p)x^*_{p(-\beta-2\alpha)}(F_3t^{-2p}u^{-p}).$$

It follows that

(2) $F_1 = F_2 = 0$.

Put $z = w^*_\alpha(t)\ w^*_\alpha(-t)$. If $p = 3$ then $z = 1$. If $p = 2$ then $z \in Z(T^*) \cap R_u$. Anyway

(3) $\mathrm{Int}\ (w^*_\alpha(t))\ x^*_{\beta+3\alpha}(-t^3u) =$

$\mathrm{Int}\ (zw^*_\alpha(-t)^{-1})\ x^*_{\beta+3\alpha}(-t^3u) = x^*_\beta(u)$ because $F_1 = 0$.

Now let both sides of (1) act on $x^*_\beta(u)$. One gets

$x^*_{2\beta+3\alpha}(\pm t^3u^2) = x^*_{2\beta+3\alpha}(\pm t^3u)\ x^*_{p(\beta+\alpha)}(\pm F_3t^pu^p)$, whence

(4) $F_3 = 0$.

We see from (2), (4) that Int $(w^*_\alpha(t))$ maps $h^*_\beta(u)$ to $h^*_{\beta+3\alpha}(\pm t^3 u)$. (The signs that are involved can be calculated in G, where the corresponding relation holds. See [22], Lemma 20). From (3) it is seen that we have the same situation with β replaced by $\beta + 3\alpha$. Hence $w^*_\alpha(t)$ normalizes the torus that is generated by the elements $h^*_\beta(u)$, $h^*_{\beta+3\alpha}(u)$. But that is T^*, so we are done for (ii) in case (a). In case (b) we skip the proof of (2), note that $w^*_\alpha(t)$ normalizes ker $(\alpha : T^* \to K)$ and obtain the same result. So we have proved (ii).

Next we prove (i). Consider $(h^*_\beta(t), w^*_\alpha(1))$ where β is a long root with $<\alpha,\beta> = 1$. This commutator is an element of T^* and is also equal to

$w^*_\alpha(t)\ w^*_\alpha(1)^{-1} = h^*_\alpha(t)$.

Finally we prove (iii).

As Int $(w^*_\alpha(t))\ x^*_{-\alpha}(u) \in \theta(\underline{g}^*_\alpha + \underline{g}^*_{p\alpha})$ (see [2], (3,3)Remark 1) while Int $(w_\alpha(t))\ x_{-\alpha}(-t^{-1}) = x_\alpha(t)$, we have Int $(w^*_\alpha(t))\ x^*_{-\alpha}(-t^{-1}) =$

$= x^*_\alpha(t)\ x^*_{p\alpha}(At^p)$, $A \in K$. So $w^*_{-\alpha}(-t^{-1}) = (x^*_\alpha(t))^{-1} w^*_\alpha(t)\ x^*_{-\alpha}(-t^{-1}) =$

$(x^*_\alpha(t))^{-1}\ x^*_\alpha(t)\ x^*_{p\alpha}(At^p)\ w^*_\alpha(t) = x^*_{p\alpha}(At^p)\ w^*_\alpha(t)$. Or

(5) $\quad w^*_{-\alpha}(-t^{-1}) = x^*_{p\alpha}(At^p)\ w^*_\alpha(t)$.

Now take a long root β such that $<\alpha,\beta> = -1$ and put $\gamma = \beta - <\beta,\alpha>\alpha$.

One has Int $(w^*_\alpha(t))\ x^*_\beta(u) = x^*_\gamma(\ldots) =$

Int $((x^*_\alpha(t))^{-1})\ x^*_\gamma(\ldots) =$ Int $(x^*_{-\alpha}(-t^{-1})\ x^*_\alpha(t))\ x^*_\beta(u) =$

$=$ Int $(w^*_{-\alpha}(-t^{-1}))\ x^*_\beta(u)$.

So Int $(w^*_{-\alpha}(-t^{-1}))\ x^*_\beta(u) = x^*_\gamma(\ldots)$, and hence the present value of c_α is just the value that makes that $w^*_{-\alpha}(-t^{-1})$ normalizes T^* (see Proof of (i) and note that we are in case (a) or (b)). Then we see from (5) that $x^*_{p\alpha}(At^p)$ normalizes T^*. So $x^*_{p\alpha}(At^p) = 1$.

12.6. In 12.5 we have seen how for every choice of c_α there is a natural choice for $c_{-\alpha}$. There still remains much freedom of choice, which we shall use to get nice actions of the $w^*_\beta(t)$ on the $x^*_\alpha(u)$ (α short, β long).

PROPOSITION. (P3).

Let R_u be commutative. There is a choice for the values (in k) of the norming constants, such that

(i) For each short root α the three statements of 12.5 hold.

(ii) For each long root β and each nonzero weight γ of $\underline{g}^*$ one has $\mathrm{Int}\,(w^*_\beta(t))\, x^*_\gamma(u) = x^*_{\gamma-<\gamma,\beta>\beta}(\pm t^{-<\gamma,\beta>}u)$.

(iii) If G is of type F_4 then $c_{2,2,\alpha_3,\alpha_4} = 0$.

PROOF.

The relation in (ii) is satisfied if $\alpha = \gamma - <\gamma,\beta>\beta$ is not a short root (use that $\mathrm{Int}(w^*_\beta(t))\, G^*_\gamma \subset G^*_\alpha$). If α is short, then γ is also a short root and $\mathrm{Int}(w^*_\beta(t))x^*_\gamma(u) = x^*_\alpha(\pm t^{-<\gamma,\beta>}u)x^*_{p\alpha}(C_{\gamma,\alpha}t^{-p<\gamma,\beta>}u^p)$, where $C_{\gamma,\alpha} \in K$ (cf. proof of 12.5). The value of $C_{\gamma,\alpha}$ depends on c_α and c_γ. Fixing c_α (or c_γ) a suitable choice of the other one kills $C_{\gamma,\alpha}$. We want to kill all $C_{\alpha,\gamma}$ simultaneously.

a) First consider case B_1. Fix c_{ε_1} and choose c_{ε_i} such that $C_{\varepsilon_1,\varepsilon_i} = 0$. Choose $c_{-\varepsilon_i}$ as indicated in 12.5. We have to prove that this is compatible with the requirements $C_{\pm\varepsilon_i,\pm\varepsilon_j} = 0$ ($i \neq j$). First we note that it follows from $w^*_\beta(t)^{-1} = w^*_\beta(-t)$ (β long) that $C_{\varepsilon_i,\varepsilon_1} = C_{\varepsilon_1,\varepsilon_i} = 0$. Next it follows from the action of $w^*_{\varepsilon_1-\varepsilon_j}(t)$ that $C_{\varepsilon_1,\varepsilon_i} = C_{\varepsilon_1,\varepsilon_j} = 0$ implies $C_{\varepsilon_j,\varepsilon_i} = 0$. The remainder then follows from the action of the elements $w^*_{\varepsilon_i}(t)$ (see Proposition 12.5, (ii), (iii)).

b) Next consider case F_4. The subgroup W_1 of W generated by reflections with respect to long roots has three orbits of degenerate sums.

(Compare with the three orbits of degenerate sums in type D_4. See 2.8). Each of these orbits can be handled like case B_1. In case B_1 we started with fixing c_{ε_1}. Now we start with fixing c_{α_3}, c_{α_4}, $c_{\alpha_3+\alpha_4}$ in such a way that (iii) holds. This can be done because α_3, α_4, $\alpha_3+\alpha_4$ lie in distinct orbits of W_1.

c) Finally consider case G_2. We use the same notations for the roots as in 10.15. Fix c_α and choose c_β, c_γ such that $C_{\alpha,\beta} = C_{\alpha,\gamma} = 0$. Then it follows from the action of $w^*_{\alpha-\beta}(t)$ that $C_{\beta,\gamma} = 0$. As in case B_1 we see that $C_{\beta,\alpha} = C_{\gamma,\alpha} = C_{\gamma,\beta} = 0$. After choosing $c_{-\alpha}$, $c_{-\beta}$, $c_{-\gamma}$ as in 12.5 we know that both $w^*_\alpha(t)$ and Int $(w^*_{\alpha-\beta}(u))\, w^*_\beta(t)$ normalize T^*. Comparing these two elements it is easy to see that $C_{-\alpha,-\beta} = 0$ (use that in G the relation Int $(w_{\alpha-\beta}(u))\, w_\beta(t) = w_\alpha(\pm tu)$ holds).

REMARKS.

1) In Proposition 12.6 the norming constant c_α may be prescribed for the short simple roots α. Then all other norming constants are fixed (see the proof of 12.6).

2) If G is of type B_3, $p = 2$, and $c_{2,2,-\varepsilon_1-\varepsilon_2,\ \varepsilon_1+\varepsilon_2+\varepsilon_3} \neq 0$ then Int $(w^*_{\varepsilon_1+\varepsilon_2}(t))\, x^*_{\varepsilon_3}(u) \neq x^*_{\varepsilon_3}(u)$.

§13. The Theorem of generators and relations and its consequences.

In this section we shall give a description of G^* in terms of generators and relations, assuming that the radical is commutative. As a result we shall get a uniqueness theorem.

13.1. Let R_u be commutative. Then we norm the homomorphism $\tau : \underline{r}_u \to R_u$ (see Theorem 11.27) such that $\tau(uZ^*_\gamma) = x^*_\gamma(u)$ for γ degenerate.

NOTATION. The kernel of τ is denoted Q. This is a finite group (see Theorem 11.27).

13.2. THEOREM. (Generators and relations).

Let R_u be commutative and nontrivial (cf. 11.4).

(i) G^* has generators $x^*_\alpha(t)$, $\alpha \in \Sigma$ or α degenerate and $t \in K$, with defining relations:

(A) If α is not a short root, then

$x^*_\alpha(u)\ x^*_\alpha(v) = x^*_\alpha(u+v)$.

If α is a short root, then

$x^*_\alpha(u)\ x^*_\alpha(v) = x^*_\alpha(u+v)\ x^*_{p\alpha}(\varepsilon_\alpha f(u,v))$, where $\varepsilon_\alpha = \pm 1$ and f is a Witt-cocycle (see 11.15).

(B) If $\alpha,\beta \in \Sigma$, $\alpha+\beta \neq 0$, then

$$(x^*_\alpha(t),\ x^*_\beta(u)) = \prod_{i>0,j>0} x^*_{i\alpha+j\beta}\ (c_{ij\alpha\beta}\ u^i v^j),$$

where the product is taken in some order and $c_{ij\alpha\beta} \in k$.

(C) $h^*_\alpha(tu) = h^*_\alpha(t)\ h^*_\alpha(u)$ for $\alpha \in \Sigma$, $t,u \in K^\times$.

Here $h^*_\alpha(t) = x^*_\alpha(t)\ x^*_{-\alpha}(-t^{-1})\ x^*_\alpha(t)\ x^*_\alpha(1)^{-1}\ x^*_{-\alpha}(-1)^{-1}\ x^*_\alpha(1)^{-1}$.

(D) There is a map $\tau' : \underline{r}_u \to G^*$ satisfying

(D1) τ' is a homomorphism of abstract groups,

(D2) $\tau'(uZ^*_\gamma) = x^*_\gamma(u)$ for γ degenerate, $u \in K$.

(D3) $\mathrm{Int}\ (x^*_\alpha(t))\ \tau'(X) = \tau'(\widehat{\mathrm{Ad}}(x_\alpha(t))X)$ for $\alpha \in \Sigma$, $X \in \underline{r}_u$, $t \in K$.

(D4) $\tau'(Q) = 1$.

(ii) Given the order of the products in (B) the values of the constants ε_α, $c_{ij\alpha\beta}$ only depend on G and on the choice of the elements X^*_α, Z^*_γ in $\underline{g}^*$ (see Theorem 3.5).

(iii) If relation (D4) is omitted, then the result is an abstract group that contains $\tau(Q)$ as a finite central subgroup.

REMARKS.

1) In order to get relations in terms of the generators $x^*_\alpha(t)$

one has to express the elements $\tau'(X)$ ($X \in \underline{r}_u$) explicitly in terms of those generators. This can be done with (D1), (D2), (D3), because the elements Z^*_γ generate $\underline{r}_u$ as a G-module.
2) The generators $x^*_\alpha(t)$ have been chosen as in Proposition 12.6 in order to fix the constants $c_{ij\alpha\beta}$. Part (ii) of the Theorem should be understood correspondingly.

PROOF.

(i) We know that these relations hold in G^* (Choose $\tau' = \tau$). We have to prove that they are defining relations. So let H be the abstract group defined by them. Then $\tau'(\underline{r}_u)$ is a normal subgroup of H (see (D1), (D3)), so we can form $H/\tau'(\underline{r}_u)$. It is easily seen that $H/\tau'(\underline{r}_u)$ satisfies Steinbergs defining relations for G (see [23] and recall that G is simply connected by Lemma 7.1). We choose a set theoretical section s of $H \to G$, with $s(1) = 1$. Every element of H can be written in the form $\tau'(X)\, s(x)$, $X \in \underline{r}_u$, $x \in G$. If this element is projected onto $1 \in G^*$, then $x = 1$, $\tau(X) = 1$ in G^*, and hence $X \in Q$. But then $\tau'(X) = 1$ in H too (see (D4)). We see that $H \to G^*$ is bijective.

(ii) We already know that the constants $c_{ij\alpha\beta}$ don't depend on G^* if they are not of the form $c_{pi,pj,\alpha,\beta}$. The constants ε_α are obtained from Proposition 11.15 (iii) (cf. Proposition 6.2). So we have only to consider the constants $c_{pi,pj,\alpha,\beta}$. It easily follows from 2.8 and from Proposition 2.12 that there are essentially four possibilities (cf. proof of 12.5).

a) α,β are simple roots in G_2 and α is the short one.

b) α,β are short roots in G_2, making an angle $2\pi/3$.

c) α,β are simple roots in a subsystem of type B_2 and α is the short one.

d) α,β are short roots in F_4, making an angle $2\pi/3$.

In case a) the constant $c_{pp\beta\alpha} = 0$, as can be seen from the relation Int $(w^*_{-\beta}(t))\ x^*_\alpha(u) = x^*_{\alpha+\beta}\ (\pm\ t^{-1}u)$. Then $c_{p,p,\beta+3\alpha,-\alpha}$ is also zero, of course. Now $c_{2p,p,\alpha,\beta}$ can be determined from the relation Int $(w^*_{-\alpha}(t))\ x^*_\beta(u) = x^*_{\beta+3\alpha}(\pm t^{-3}u)$. (Its value depends on the order. Use $(x,y) = (y,x)^{-1}$.)

Once we know the values of $c_{p,2p,\beta,\alpha}$ and $c_{p,2p,\beta+3\alpha,-\alpha}$ we can determine $c_{p,p,-\alpha,\beta+2\alpha}$ from the same relation. This will do in case a) and b).

In case c) we argue as in case a) and see that $c_{pp\beta\alpha} = 0$.

Finally consider case d). One of the constants of this type is known to be zero: $c_{2,2,\alpha_3,\alpha_4} = 0$ (see 12.6). It is seen from the relation $(x^*_\gamma(t),\ x^*_\delta(u))^{-1} = (x^*_\delta(u),\ x^*_\gamma(t))$ $(\gamma = \alpha_3,\ \delta = \alpha_4)$ that $c_{2,2,\delta,\gamma} = 1$. The constant $c_{2,2,-\delta,\gamma+\delta}$ can be determined from the relation

Int $(w^*_{-\delta}(t))\ x^*_\gamma(u) = x^*_{\gamma+\delta}\ (tu)\ x^*_{2\gamma+2\delta}\ (\ldots)$.

In the same way all $c_{2,2,\alpha,\beta}$ can be found with α,β lying in the plane through γ,δ. We now need the following Lemma:

13.3. LEMMA.

Let Σ be of type F_4. The subgroup W_1 of W generated by reflections with respect to long roots acts transitively on the planes spanned by pairs of short roots, making an angle $2\pi/3$.

PROOF.

Let S be the set of such planes. There are three orbits of short roots under the action of W_1 (see proof of 12.6). It is seen from the explicit form of these orbits that

(1) If α,α' are short roots in the same orbit, then $\alpha = \pm\alpha'$ or $(\alpha,\alpha') = 0$.

If $V \in S$ then V contains a representative of each of the three orbits. Let $V' \in S$. We have to prove that there is $w \in W_1$ with $wV = V'$. We may assume that $V \cap V'$ contains a root α. Let $\beta \in V$, $\beta' \in V'$ be short roots with

(2) $\langle\alpha,\beta\rangle = \langle\alpha,\beta'\rangle = -1$.

If β,β' ly in distinct orbits, then we replace β by $-\alpha-\beta$, which lies in the same orbit as β' (use (1)). If $\beta = \pm\beta'$ then $V = V'$ and we are done. So we may assume $(\beta,\beta') = 0$ (see (1)). Then $\beta-\beta'$ is a long root and we use the reflection with respect to $\beta-\beta'$. It follows from (2) that $(\beta-\beta',\alpha) = 0$, and we see that V' is transported to V.

13.4. PROOF CONTINUED.

From the Lemma it follows that all $c_{2,2,\alpha,\beta}$ in case d can be derived from those in the plane through α_3, α_4 by means of the actions Int $(w_\zeta^*(t))$ with ζ long. This finishes the proof of (ii). Part (iii) is an easy consequence of the fact that Q is fixed by G.

13.5. COROLLARY.

Let $\phi: G^* \to G$ and Q be as above with commutative radical R_u (see 13.1, 11.4). Let $0 \to \underline{r}_u \xrightarrow{\tau_1} G_1^* \xrightarrow{\phi_1} G \to 1$ be the extension from Theorem 10.1. Then there is a separable k-homomorphism χ from G_1^* onto G^* such that

(i) The kernel of χ is $\tau_1(Q)$,

(ii) $\phi \circ \chi = \phi_1$.

REMARK. We don't claim that χ is unique.

PROOF.

From the Theorem it follows that there is a homomorphism χ of abstract groups, sending $x^*_{\alpha 1}(u)$ to $x^*_{\alpha}(u)$, where $x^*_{\alpha 1}(u)$ is the a analogue of $x^*_{\alpha}(u)$. One argues as in the end of the proof of theorem 9.6 to see that χ is a morphism. On the open cell χ is defined over k, so χ is defined over k (see [19], Lemma 1).

13.6. NOTATION. If H is an algebraic group, then Aut(H) denotes the abstract group of automorphisms (in the sense of algebraic groups) of H.

13.7. COROLLARY.

Let ϕ: $G^* \to G$ be given as in 13.5.

(i) The natural homomorphism $\mathrm{Aut}(G^*) \to \mathrm{Aut}(G)$ is surjective.

(ii) Aut (G^*) can be given the structure of an algebraic group with $\dim(\mathrm{Aut}(G^*)) = \dim G^*$.

PROOF.

(i) Let $\psi: G \to G$ be an automorphism. We have to show that there is χ: $G^* \xrightarrow{\sim} G^*$ with $\phi \circ \chi = \psi \circ \phi$.

If ψ is inner then it is easy. So we assume ψ to be a graph automorphism (see [22], p. 157). We have $\psi(x_\alpha(t)) = x_{\sigma\alpha}(\varepsilon'_\alpha t)$, where σ is the permutation of Σ corresponding to ψ and $\varepsilon'_\alpha = \pm 1$. As we only consider automorphisms of algebraic groups we only have to do with the case that σ preserves root lengths. If $Q = 0$ then it is easy to see that $x^*_\alpha(t) \mapsto x^*_{\sigma\alpha}(\varepsilon'_\alpha t)$ preserves relations (A), (B), (C), (D). If $Q \neq 0$ then Σ is of type B_1 or G_2 (see 11.27, 5.2). But then σ is trivial. It is seen as in the proof of 13.5 that $x^*_\alpha(t) \mapsto x^*_{\sigma\alpha}(\varepsilon'_\alpha t)$ defines an automorphism of algebraic groups.

(ii) First assume $Q = 0$.

Put $N = \ker(\mathrm{Aut}(G^*)) \to \mathrm{Aut}(G)$. If $\chi \in N$, then χ can be written in the form $\chi_1 \circ \chi_2$, where $\chi_1 = \mathrm{Int}(x)$ for some $x \in R_u$, $\chi_2(T^*) = T^*$ (use that maximal tori are conjugate in $T^* . R_u$). Say $\chi = \chi_2$. Then T^* is fixed by χ, because $\chi \in N$. So $\chi(x^*_\alpha(t)) = x'_\alpha(t)$, where $x'_\alpha(t)$ is obtained by replacing the norming constants c_α by constants c'_α (use Lemma 11.16). As the $x'_\alpha(t)$ satisfy relations (A), (B), (C), (D) the values of the c'_α are determined by the values for α short and simple (see 12.6, Remark 1).

We claim that these values can be obtained from an inner automorphism in the group $\ulcorner C,G \urcorner$ that is discussed in section 10 (cf.10.21).

Proof of the claim: Put $H = N_{\ulcorner C,G \urcorner} G^*/Z_{\ulcorner C,G \urcorner} G^*$ (cf.10.21). Then H acts on G^* in a natural way and G^* also acts on H. The unipotent radical of H can be viewed as a G^*-module M, with $\dim M \geqslant \dim \underline{r}_u$ (use 10.22 and the structure of $\ulcorner C,1 \urcorner$ as a G^*-module). The homomorphism of abstract groups $H \to \mathrm{Aut}(G^*)$ maps M into N. There is a natural homomorphism $\rho: \underline{r}_u \to M$. For each $x \in M$ there is $X \in \underline{r}_u$ such that $x\rho(X)$ fixes T^*. It easily follows that $\dim M_0 \geqslant \dim(\underline{r}_u)_0$. But $\dim(\underline{r}_u)_0$ is equal to the number of short simple roots (see Proposition 5.2), whence the claim. (Use that $c'_\alpha - c_\alpha$ depends linearly on $m \in M_0$). It also follows that $\dim M_0 = \dim(\underline{r}_u)_0$, so $\dim H = \dim G^*$. Summing up we conclude that N is contained in the image of H, and that $\dim H = \dim G^*$. It is easy to see now that H is isomorphic (as an abstract group) to the inverse image in $\mathrm{Aut}(G^*)$ of the normal subgroup $\mathrm{Int}(G)$ of $\mathrm{Aut}(G)$. The finite subgroup F of graph automorphisms in $\mathrm{Aut}(G)$ (that satisfy $\varepsilon'_\alpha = \varepsilon'_{-\alpha} = 1$ for α simple) can be lifted to $\mathrm{Aut}(G^*)$ (see proof of (i)). We see that $\mathrm{Aut}(G^*)$ is isomorphic as an abstract group to the semi-direct product of H and F (see [1], (1.11)). If $F \neq 1$ then $Z(G^*) = 1$, so $H \cong G^*$.

Hence Aut(G^*) can be given the structure of an algebraic group with dim (Aut(G^*)) = dim H = dim G^*.
If Q is nonzero then we see from the proof of (i) that $\tau_1(Q)$ is fixed by any element of Aut(G_1^*), where G_1, τ_1 are as in Corollary 13.5 (use that $\ulcorner C,1 \urcorner$ commutes with $\tau_1(Q)$). So Aut(G^*) $\cong$ Aut(G_1^*).

13.8. THEOREM. (Uniqueness).
Let $\phi: G^* \to G$, $\phi': G^{*\prime} \to G$ be two solutions of $d\phi = \pi$ with commutative radicals (see 7.2). Let Q, Q' be corresponding subgroups of $\underline{r}_u$ (see 13.1). Then the following statements are equivalent

(i) Q = Q'.

(ii) G^* is isomorphic to $G^{*\prime}$.

(iii) There is an isomorphism $\chi: G^* \to G^{*\prime}$ such that $\phi' \circ \chi = \phi$.

PROOF.
(ii) follows from (iii).
(iii) follows from (i) by Corollary 13.5 (note that a separable surjective homomorphism is a quotient morphism in the sense of [1], Ch. II, § 6).
We still have to prove that (i) follows from (ii). The isomorphism $\chi: G^* \to G^{*\prime}$ induces an isomorphism $\rho: G \to G$ with $\phi' \circ \chi = \rho \circ \phi$ (use that ϕ, ϕ' both "divide out" the radicals). From Corollary 13.7 (i) it follows that we may assume ρ to be the identity. Then we change χ by an inner automorphism Int(x), $x \in R_u$, such that $\chi(T^*) = T^{*\prime}$. The homomorphisms $\tau': \underline{r}_u \to G^{*\prime}$ and $\chi \circ \tau: \underline{r}_u \to G^{*\prime}$ then coincide, because $d\chi = id: \underline{g}^* \to \underline{g}^*$ (use the universal property of $\pi: \underline{g}^* \to \underline{g}$). So Q = ker τ = ker $(\chi \circ \tau)$ = ker τ' = Q'.

13.9. THEOREM.

(i) Let $\phi: H \to G$ be given as in 11.1 such that (P2) holds (see 11.1). Assume that G is simply connected. Let $0 \to \underline{r}_u \xrightarrow{\tau_1} G_1^* \xrightarrow{\phi_1} G \to 1$ be the extension from Theorem 10.1. Then there is a k-homomorphism χ from G_1^* into H such that $\phi \circ \chi = \phi_1$.

(ii) Let G be a semi-simple algebraic group with perfect Lie algebra (cf. proof of 11.30). If $p = 2$ assume that G has no factor of type B_3. Then there is a connected linear algebraic group G_1^* and a homomorphism ϕ_1: $G_1^* \to G$ such that:

(a) ϕ_1 is an infinitesimally central extension and $\underline{g}_1^* = [\underline{g}_1^*, \underline{g}_1^*]$.

(b) If H is a connected linear algebraic group with $\underline{h} = [\underline{h}, \underline{h}]$ and ϕ: $H \to G$ is an infinitesimally central extension, then there is a surjective separable homomorphism χ: $G_1^* \to H$ such that $\phi \circ \chi = \phi_1$. If χ' : $G_1^* \to H$ also satisfies $\phi \circ \chi' = \phi_1$ then there is an automorphism ξ of G_1^* such that $\chi = \chi' \circ \xi$.

(c) $d\phi_1$ is a universal central extension.

PROOF.

(i) As τ is H-equivariant, $\tau(\underline{r}_u)$ is a normal subgroup. Put $H' = H/\tau(\underline{r}_u)$ and let ϕ': $H' \to G$ be the homomorphism induced by ϕ. Then ϕ' satisfies (P2) in a trivial way and hence Steinbergs relations (A), (B) hold in H' (see section 11). It follows from ([23], Théorème 3.3) that relation (C) is satisfied for arguments that are algebraic over the prime field. Then relation (C) holds for all arguments for reasons of continuity. It follows (cf. proof of Corollary 13.5) that ϕ' splits, i.e. there is a homomorphism ψ: $G \to H'$ such that $\phi' \circ \psi = id$ (use that G is simply connected). We may replace H by the inverse image of $\psi(G)$ in H. Then ϕ: $H \to G$ is still of the type described in 11.1 and (P3) holds (cf. 11.4; use Lemma 11.16 for separability). If follows from Lemma 11.2 that $d\phi$ is a central extension. Hence there is a homomorphism of Lie algebras $\rho: \underline{g}^* \to \underline{h}$ such that $d\phi \circ \rho = \pi$. From the central trick it follows that ρ is H-equivariant, where H acts on $\underline{g}^*$ by $\hat{Ad} \circ \phi$. It is seen from the structure of $\underline{r}_u$ as a G-module (H-module) that $(d\tau)(\underline{r}_u)$ is the direct sum of $\rho(\underline{r}_u)$ and an H-submodule $\underline{c}$. So $\underline{h}$ is the direct sum of $\rho(\underline{g}^*)$ and $\underline{c}$. The action of H on $\underline{h}$ factors over G (see Lemma 11.2). Now we use

13.10. LEMMA.

Let ϕ: H → G be given as in 11.1, such that (P3) holds. Then H has generators and relations like those in Theorem 13.2, with constants ε_α, $c_{ij\alpha\beta}$ that only depend on G, the action of G on $\underline{h}$ and the choice of the elements X^*_α, Z^*_γ in $\underline{h}$ (defined as indicated in 11.3).

REMARK. The group Q (= ker τ) corresponding to H is not necessarily finite.

The proof of the Lemma is the same as that of Theorem 13.2.

13.11. We continue the proof of Theorem 13.9, (i). Consider the semi-direct product of $\underline{c}$ and $G^*_1/\tau_1(\ker \rho)$, where G^*_1 is as in the Theorem. This is a group S with the same Lie algebra as H and with the same action of G on that Lie algebra. Then it follows from Lemma 13.10 (cf. Corollary 13.5) that there is a homomorphism χ': S → H such that its composition χ with the natural homomorphism $G^*_1 \to S$ satisfies $\phi \circ \chi = \phi_1$ (k-rationality follows as in 13.5).

(ii) As $\underline{g}$ is perfect, the simply connected covering $G^{sc} \to G$ is separable (see proof of Lemma 7.1). Each almost simple factor G^{sc}_i of G^{sc} has an extension ϕ_i as in Theorem 10.1. The direct product of these extensions is an extension $\phi^{sc}: G_1^* \to G^{sc}$ such that $d\phi^{sc}$ is a universal central extension. We get an extension $\phi_1: G_1^* \to G$ from it, such that $d\phi_1$ is a universal central extension (use that $G^{sc} \xrightarrow{\psi} G$ is separable). Now assume $\phi: H \to G$ is given such that $d\phi$ is a central extension and such that $\underline{h}$ is perfect. Let G_i be an almost simple factor of G, T^* a maximal torus in H and T_i^* a subtorus of T^* such that $\phi(T_i^*)$ is a maximal torus T_i in G_i (cf. proof of Theorem 11.30). There is a surjective homomorphism of Lie algebras $\rho: \underline{g}_1^* \to \underline{h}$ such that $d\phi \circ \rho = d\phi_1$ (see Proposition 1.3, (v)). It is H-equivariant (use the central trick). Consider the composite homomorphism $H \to G \to G_i$ and the tori T_i^*, T_i. The situation is that of 11.1 with (P1) (cf. proof of Theorem 11.30). If G_i is not simply connected then it follows as in the proof of (i) that there is a homomorphism χ_i from G_i^{sc} into H, such that $\phi \circ \chi_i = \psi_i$. If G_i is simply connected, then it follows from Theorem 11.30, Corollary 11.29, Remark 2 in 11.1, that (P2) holds. So we can apply (i). The result is a homomorphism $\chi: G_1^* \to H$ such that $\phi \circ \chi = \phi_1$ (use Lemma 7.1). Then $d\chi = \rho$, because $d\phi \circ d\chi = d\phi_1$. So χ is surjective and separable, which proves the existence of χ in (b). Now suppose $\chi' : G_1^* \to H$ also satisfies $\phi \circ \chi' = \phi_1$. Let T_1^* denote a maximal torus of G_1^* such that $\chi(T_1^*) = T^*$. We may change χ' by an automorphism $\mathrm{Int}(x)$, $x \in R_u(G_1^*)$, such that $\chi'(T_1^*) = T^*$. We have morphisms $x_{\alpha,i}^* : K \to G_1^*$ as in the proof of Theorem 11.30. As $H \to G_i$ satisifes (P1) (see above), we may apply Lemma 11.16 to see that χ,χ' coincide on $x_{\alpha,i}^*(t)$ if α is a long root with respect to G_i. Furthermore we can "change the norming constants" by an automorphism ξ such that $\chi' \circ \xi$ and χ also coincide on $x_{\alpha,i}^*(t)$ for α short and simple (see proof of 13.7). Then $\chi' \circ \xi = \chi$ because they coincide on generators (cf. 12.6,

Remark 1). Parts (a), (c) follow from the construction above.

13.12. We return to the notations of 11.4.

COROLLARY.

Let M be an indecomposable nonzero quotient of the G-module $\underline{r}_u$. Then $\dim_k H_k^2(G,M) = 1$.

PROOF.

By Theorem 13.9 (i) an extension of G by M is either isomorphic to a quotient of the extension from Theorem 10.1 or it splits. So there is only one nontrivial 2-cocycle, up to scalar multiples.

13.13. PROPOSITION.

Let M be a G-module in which all nonzero weights are degenerate sums. Let $\overline{f} \in H^2(G,M)$. Then there is a homomorphism of G-modules $\rho: \underline{r}_u \to M$ such that $\overline{f}$ is in the image of $H^2(\rho): H^2(\underline{r}_u) \to H^2(M)$.

PROOF. Consider the extension $\phi : H \to G$, corresponding to $\overline{f}$. The weights of M lie in $p\Gamma$ but the roots do not, so the differential of the action of G on M is trivial (Use [2], Lemma 5.2). So $d\phi$ is a central extension and there is a homomorphism of Lie algebras $\rho:\underline{g}^* \to \underline{h}$ such that $d\phi \circ \rho = \pi$. We claim that the restriction of ρ to $\underline{r}_u$ satisfies the requirements. It is sufficient to prove that the image of $\overline{f}$ in $H^2(M/\rho(\underline{r}_u))$ is zero, because the case of $H^2(\rho(\underline{r}_u))$ is discussed in Theorem 13.9 (i) (use Lemma 11.16 to prove linearity of the restriction of χ to $\underline{r}_u$ in 13.9 (i)). So we may assume that $d\phi$ splits (replace M by $M/\rho(\underline{r}_u)$). In this case we prove that $\overline{f}$ is trivial by induction on the number of irreducible factors of M. If M is irreducible then the result follows from Theorem 13.9 (i) or Theorem 9.6 (see Proposition 5.2 and classify M by its highest weight). If $0 \to L \to M \to N \to 0$ is an exact sequence of G-modules, $L \neq 0$, then $H^2(L) \to H^2(M) \to H^2(N)$ is exact, and the image of $\overline{f}$ in $H^2(N)$ is zero by induction hypothesis. So $\overline{f}$ is the image of some $\overline{g} \in H^2(L)$, which is zero by the same reason. (A sub-extension of a splitting central extension splits by the central trick).

13.14. THEOREM.

Let $\tilde{G}$ be a simply connected almost simple subgroup of G. Assume there is a long root α (with respect to G,T) such that $X_\alpha \in \tilde{\underline{g}}$, $h_\alpha(t) \in \tilde{G}$ for $t \in K^\times$. Assume furthermore that $\tilde{T} = \tilde{G} \cap T$ is a maximal torus in $\tilde{G}$.
Let $\tilde{\underline{g}}$ be perfect and let $0 \to \underline{r}_u \to G^* \xrightarrow{\phi} G \to 1$, $0 \to \tilde{\underline{r}}_u \to \tilde{G}^* \xrightarrow{\tilde{\phi}} \tilde{G} \to 1$ be the extensions from Theorem 10.1.
Then there is a homomorphism $\psi: \tilde{G}^* \to G^*$ such that $\phi \circ \psi = \tilde{\phi}$.

REMARK. Again we don't claim that ψ is unique.

PROOF.

There is a dual pairing $X(T) \times X_*(T) \to \mathbb{Z}$, where $X(T)$ is the character group of T and $X_*(T)$ is the group of one parameter subgroups of T (see [1], (8.6)). Note that $X(T)$ is just Γ. We denote the pairing <,>, as in loc. cit. There are natural maps $X(T) \to X(\tilde{T})$ and $X_*(\tilde{T}) \to X_*(T)$. Let V be the real vector space in which Σ, Γ are imbedded. There is a natural choice for the inner product (,) on V and on its dual V', up to scalar factors. This choice is characterized by the fact that (,) is invariant under W (see [4], Ch. VI, § 1, n° 1.2, Proposition 7). We choose (,) in the following way:
For $\lambda,\mu \in X_*(T)$, we put

$$(\lambda,\mu) = \sum_{\gamma \text{ weight of } \underline{g}} \langle\gamma,\lambda\rangle\langle\gamma,\mu\rangle,$$ extend this to V', and identify

V with V' by means of this inner product. Then we restrict (,) to the subspace $\tilde{V}'$ spanned by $X_*(\tilde{T})$, which we view as a subset of $X_*(T)$ (cf. [14], §2). We get an inner product that is invariant under the Weyl group $\tilde{W}$ of $\tilde{G}$ (use that $\underline{g}$ is a $\tilde{G}$-module). Then we identify $X(\tilde{T}) = \tilde{\Gamma}$ with a subset of $\tilde{V}'$ by means of the inner product. The result is that we have embeddings of $X_*(T)$, $X_*(\tilde{T})$, $X(T)$, $X(\tilde{T})$ into a real vector space V with inner product (,). In V the map $X(T) \to X(\tilde{T})$ corresponds to the orthogonal projection of V on the subspace $\tilde{V}$ (or $\tilde{V}'$). The long root $\alpha \in \Sigma$ is its own projection because $t \mapsto h_\alpha(t)$ is in $\tilde{V}$. If γ is a degenerate sum in Γ, then its projection on $\tilde{V}$ is an element $\tilde{\gamma}$ of $p\tilde{\Gamma}$ with $(\tilde{\gamma},\tilde{\gamma}) \leqslant (\gamma,\gamma) \leqslant p(\alpha,\alpha)$ (see Proposition 2.12). If $\tilde{\gamma} \in \tilde{\Gamma}_0$, then $\tilde{\gamma}$ is either zero or degenerate by Proposition 2.12. Consider the inverse image H of $\tilde{G}$ in G*. It is an extension of $\tilde{G}$ by $\underline{r}_u$, where the weights of $\underline{r}_u$ are zero, degenerate or not contained in $\tilde{\Gamma}_0$. Write $\underline{r}_u = M \oplus N$ where M is spanned by the weight components of weights in $\tilde{\Gamma}_0$ (cf. 10.14, Remark). We claim that $H^2(G,N) = 0$. Then the result follows from Proposition 13.13. So we still have to prove:

13.15. PROPOSITION.

_If_ N _is a_ G-_module with weights that are not in_ Γ_0, _then_ $H^2(G,N) = 0$.

PROOF.

Let $\phi: H \to G$ be an extension of G by N. From Theorem 8.2 we get the existence of a T*-equivariant cross section $s: G \to H$, where T* is a maximal torus in $\phi^{-1}(T)$ as usual. We have an "open cell" $\Omega^* = \phi^{-1}(\Omega) = N \cdot s(\Omega)$. Put $x^*_\alpha(t) = s(x_\alpha(t))$ for $\alpha \in \Sigma$. We argue as in 11.15, 11.18 to see that Steinbergs relations (A), (B) hold. It follows as in the proof of Theorem 13.9 (i) that ϕ splits.

13.16. Examples to Theorem 13.14.

1) Let $\tilde{G}$ be the subgroup G_{B_3} of G_{D_4} which we discussed in 3.11. Then $\tilde{G}$ contains all elements $x_{\pm\alpha_2}(t)$, $t \in K$, and hence $X_{\alpha_2} \in \tilde{\mathfrak{g}}$ (here α_2 is the second simple root in type D_4). The other conditions are also satisfied (see 3.11) so there is a homomorphism $G^*_{B_3} \to G^*_{D_4}$. Compare this result with the construction of $G^*_{B_3}$ in 10.12.

2) Similar examples are obtained from the "triality" in D_4 (cf. Remark 10.17) and from the graph automorphisms of G_{D_l} ($l > 4$).

3) The triality induces an embedding $G_{G_2} \to G_{D_4}$ that factors through the embedding from example 1. As a result we get an embedding $G_{G_2} \to G_{B_3}$ which also satisfies the requirements.

4) Let $\tilde{G}$ be the subgroup of G_{F_4} generated by the elements $x_{\pm\alpha_3}(t)$, $x_{\pm\alpha_4}(t)$, $t \in K$. It is a simply connected group of type A_2, but the assumption about the long root in 13.14 is false. If $p = 2$, then it is easy to see from the $[p]$-structures that there is no homomorphism ψ as in the Theorem.

§14. The group functor G^*.

In this section we discuss a group functor which has $R \mapsto \mathfrak{g}^*_R$ as a Lie algebra. We omit proofs.

14.1. We will consider contravariant functors from schemes to sets, which are sheaves on the category of schemes. Giving such a sheaf is equivalent to giving a covariant functor from rings to sets which is a sheaf (see [15] I § 2 (2.3.6)). We will identify these two sheaves.

14.2. Let G be a simply connected almost simple Chevalley group scheme that is not of type C_l ($l \geqslant 1$). Its Lie algebra is perfect and we have a universal central extension $\pi: \underline{g}^*_{\mathbb{Z}} \to \underline{g}_{\mathbb{Z}}$, inducing a universal central extension $\underline{g}^*_R \to \underline{g}_R$ for every ring R. So we have a functorial morphism, which we denote $\pi: \underline{g}^* \to \underline{g}$ (Here we drop the convention $\underline{g} = \underline{g}_K$). For p = 2, 3 we have an extension $\phi: G^* \to G \times_{\mathrm{Spec}(\mathbb{Z})} \mathrm{Spec}(\mathbb{F}_p)$ as in Theorem 10.1. It defines a functorial morphism of group functors on the category of (commutative) $\mathbb{F}_p$-algebras. We put $G^*_p(R) = G^*(R/pR)$ and $G_p(R) = G(R/pR)$. We get group functors on the category of rings. A functorial morphism $\phi_p: G^*_p \to G_p$ is induced by ϕ. It is in fact a morphism of group functors. We extend the functor G^* from section 10 to a functor on the category of rings defining the extension as the limit of the projective system, given by the diagram

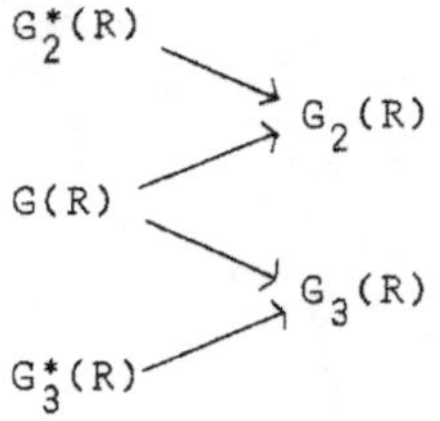

.

Equivalently, we have $G^* = (G^*_2 \times_{G_2} G) \times_{G_3} G^*_3$. It is a group functor and it is a sheaf. There is a morphism of group functors $\phi: G^* \to G$. Its kernel is isomorphic to the kernel of $\pi: \underline{g}^* \to \underline{g}$ and its differential $d\phi$ is isomorphic to π. Here the differential is taken in the sense of ([12], Exp. II, Prop. 3.7), where it is denoted $L(\phi)$. The tangent spaces may be supplied with a structure of Lie algebra functors by the definitions given in ([12], Exp. II). (One has to check a list of conditions). Then $d\phi$ may be identified with π as a homomorphism of Lie algebra functors (i.e. there are suitable isomorphisms).

14.3. If ker ϕ is nontrivial then ker ϕ (or ker π), $\underline{g}^*$, G^* are not representable by schemes. For suppose G^* is (representable by) a scheme. Then its tangent space $\underline{g}^*$ is an affine scheme (see [12], Exp. II, Prop. 3.3 and Exp. I, 4.6.3). This is not compatible with the fact that there is $x \in \underline{g}^*_{\mathbb{Z}}$, $x \neq 0$, such that its image in $\underline{g}^*_{\mathbb{F}_p}$ is zero for almost all p. In the same way we see that $\underline{g}^*$ and ker ϕ are no schemes. (They are their own tangent spaces).

References

1. A. Borel, Linear algebraic groups, W.A. Benjamin, Inc., New York (1969).

2. ——— , Seminar on algebraic groups and related finite groups A, Lecture notes in Mathematics 131, Springer, Berlin (1970), 1-55.

3. ——— and T.A. Springer, Rationality properties of linear algebraic groups II, Tôhoku Math. J., vol. 20, (1968) 443-497.

4. N. Bourbaki, Groupes et algèbres de Lie, Chap. IV, V, VI, Act. Sci. Ind., Hermann, Paris (1969).

5. H. Cartan and S. Eilenberg, Homological algebra, Princeton University Press (1956).

6. C. Chevalley, Sur certains groupes simples, Tôhoku Math. J., vol. 7, (1955), 14-66.

7. ——— , Séminaire sur la classification des groupes de Lie algébriques, Paris (1956-58).

8. ——— , Certains schémas de groupes semi-simples, Séminaire Bourbaki, 13e année, (1960-61), Exp. 219.

9. ——— , The algebraic theory of spinors, Columbia University Press, New York (1954).

10. C. Curtis and I. Reiner, Representation theory of finite groups and associative algebras, Wiley, New York (1962).

11. M. Demazure and P. Gabriel, Groupes algébriques, Tome I, Masson & cie, Paris, North-Holland, Amsterdam (1970).

12. ——— and A. Grothendieck, Séminaire de géometrie algébrique du Bois Marie, SGA3, (1962/64), Lecture notes in mathematics 151-153, Springer, Berlin (1970).

13. J.A. Dieudonné, Les algèbres de Lie simple associées aux groupes simples algébriques sur un corps de caractéristique $p > 0$, Rendiconti del Circolo Mathematico di Palermo, Serie II, tomo 6, Palermo (1957), 198-204.

14. E.B. Dynkin, Semisimple subalgebras of semisimple Lie algebras, Am. Math. Soc. Transl. Ser. 2, 6, (1957), 111-245 (= Mat. Sbornik N.S. 30 (1952), 349-462).

15. A. Grothendieck and J.A. Dieudonné, Eléments de géometrie algébrique, Die Grundlehren der Mathematischen Wissenschaften in Einzeldarstellungen, Band 166, Springer, Berlin (1971).

16. N. Jacobson, Lie algebras, Interscience Tracts in pure and applied mathematics 10, Interscience Publ., New York (1962).

17. ———, Classes of restricted Lie algebras of characteristic p, I, Am. J. 63, (1941), 481-515.

18. B. Kostant, Lie algebra cohomology and the generalized Borel-Weil theorem, Annals of Math. 74, (1961), 329-387.

19. M. Rosenlicht, Questions of rationality for solvable algebraic groups over non-perfect fields, Annali di Matematica pura ed applicata (IV) vol. 61, (1963), 97-120.

20. G.B. Seligman, Modular Lie algebras, Ergebnisse der Mathematik und ihrer Grenzgebiete, Band 40, Springer, Berlin (1967).

21. T.A. Springer, Weyl's character formula for algebraic groups, Inventiones math. 5, (1968), 85-105.

22. R. Steinberg, Lectures on Chevalley groups, Yale Univ. Lecture Notes (1967-68).

23. ———, Générateurs, relations et revêtements de groupes algébriques, Colloque sur la théorie des groupes algébriques, Bruxelles (1962), 113-127.

24. ———, Endomorphisms of linear algebraic groups, memoirs of the A.M.S. 80, (1968).

25. J. Tits, Tabellen zu den einfachen Lie gruppen und ihren Darstellungen, Lecture notes in mathematics 40, Springer, Berlin (1967).

26. F.D. Veldkamp, Representations of algebraic groups of type F_4 in characteristic 2, Journal of algebra 16, (1970), 326-339.

27. W.J. Wong, Representations of Chevalley groups in characteristic p, Nagoya Math. J. 45, (1972), 39-78.

List of Notations

We use mainly the same notations as in [1], [2], [4], [22].

$\widehat{\mathrm{Ad}}$	representation of G in $\underline{g}^*$.	3.1
$\widehat{\mathrm{ad}}$	d $\widehat{\mathrm{Ad}}$	3.3
C, (C)	G-module C, condition (C).	10.3,10.4
c_α	norming constant	11.17
$c_{ij\alpha\beta}$	constant in commutator relation	11.18
f_0, $\overline{f}_0$	element of L_2, $H^0(L_2)$.	10.3
G	Chevalley group from 2.1, except in 7.1, 7.8, 7.9, 8, 9, 11.1, 11.2, 11.3, 11.29, 11.30, 13.9, 13.10, 14. After section 4 it is assumed that if G is as in 2.1 then $\underline{g}$ is perfect (or $\Sigma \cap p\Gamma = \emptyset$).	
G^*	see ϕ.	
$G^{*\alpha}$	subgroup generated by $x^*_\alpha(t)$, $x^*_{-\alpha}(t)$, $t \in K$.	12.2
G^α	$\phi(G^{*\alpha})$.	
G^*_α	subgroup with Lie algebra $\underline{g}^*_\alpha$.	11.5
G_α	$\phi(G^*_\alpha)$.	
$G^*_{(\alpha,\beta)}$	subgroup generated by $x^*_{i\alpha+j\beta}(u)$, $i > 0$, $j > 0$.	11.19
$\underline{g}$	$\underline{g} = \underline{g}_K$	2.1
$\underline{g}^*$	see π.	
$\underline{g}'_{\mathbb{Z}}$	$r : \underline{g}'_{\mathbb{Z}} \to \underline{g}_{\mathbb{Z}}$.	2.14
H^*_α	generator of $\underline{g}^*$.	3.5
h^*_α	$h^*_\alpha(t) = w^*_\alpha(t)w^*_\alpha(1)^{-1}$.	12.2
$i_V, i_G, \ldots$	mappings into $\ulcorner V,G \urcorner$.	8.1
$\widehat{\mathrm{Int}}$	action of G on R_u.	11.26
k,K	K is algebraic closure of k.	2.1

$L_M, L_{M/N}$	G-module $M \otimes_{\mathbb{Z}} K, \ldots$	4.1
n_γ	$\max\{n \mid \gamma \in n\Gamma\}$.	3.5,3.7
(P1),...	condition or label.	11.1,11.3
$p_G, p_V, \ldots$	projections from $\ulcorner V, G \urcorner$.	8.1
Q	ker τ.	13.1
$\underline{r}_u$	G-module ker π or Lie algebra of the unipotent radical R_u of G^*.	10,7.4
T^*	torus in G^* or H.	11.1,11.4
w_α^*	$w_\alpha^*(t) = x_\alpha^*(t)x_{-\alpha}^*(-t^{-1})x_\alpha^*(t)$.	12.2
x_α^*	$x_\alpha^*(t) \in G_\alpha^*$ for $t \in K$.	11.6,11.17
X_α^*	generator of $\underline{g}^*$.	3.5
y_α^*	$x_\alpha^*(t) = y_\alpha^*(t)x_{p\alpha}^*(c_\alpha t^p)$.	11.17
Z_γ^*	generator of $\underline{g}^*$.	3.5
$Z(T^*)$	centralizer $Z_{G^*}(T^*)$ of T^* in G^*.	
$\mathcal{E}, \mathcal{E}_1, \mathcal{E}_2$	exact sequences.	9.4,10.3
$\mathcal{L}_V$	category of modules L_M, $M \subset V$.	4.1
Γ	lattice of weights.	2.1
Γ_0	sublattice generated by roots.	2.1
ε_α	$(X_\alpha^*)^{[p]} = -\varepsilon_\alpha Z_{p\alpha}^*$.	11.15
θ	morphism onto Ω^* or restriction of this morphism.	11.12
π	$\pi : \underline{g}^* \to \underline{g}$ is a u.c.e.	1.1
τ^α	morphism into $Z(T^*) \cap R_u$.	11.23
ϕ	$\phi : G^* \to G$ satisfies $d\phi = \pi$.	7
Ω^*	$\phi^{-1}(\Omega)$, where Ω is the open cell in G.	2.1,11.10

Subscripts:

$V_\gamma, \underline{g}_\gamma, V_0, \ldots$	weight spaces.
$G_{A_3}, \ldots$	G of type $A_3, \ldots$

Brackets etc.:

$\{H_\alpha\}$, $\{x\}$, $\{x\}_M$, $\{x\}_{M/N}$	residue classes.	2.16,4.1
$\ulcorner V,G \urcorner$	semi-direct product.	8.1
$\ulcorner V,\underline{g} \urcorner$,...	Lie algebra of $\ulcorner V,G \urcorner$,...	
$v_1/v_2/v_3$	generators of composition series.	4.14
$\alpha \perp \beta$	$(\alpha,\beta) = 0$.	
$R^\times$	group of invertible elements of R.	

Index

admissible	4.1
central trick	1.2
centrally closed	1.1
coboundary	9.1
cochain	9.1
cocycle	9.1
Σ-connected	4.12
degenerate sum	2.4
equivariant	8.1
extension (of Lie algebra)	1.1
(universal) central extension	1.1
extension (of group)	8.1
k-extension	8.1
infinitesimally central extension	7.8
Hochschild group	9.1
homomorphism	conventions
indecomposable	4.8
indecomposable component	4.10
Jacobi relation	1.1
Lie algebra	1.1
long root	conventions
morphism	conventions
norming constant	11.17
perfect	11.30
ring	1.1
short root	conventions
standard lattice	4.1
Witt-cocycle	11.15

Vol. 215: P. Antonelli, D. Burghelea and P. J. Kahn, The Concordance-Homotopy Groups of Geometric Automorphism Groups. X, 140 pages. 1971. DM 16,-

Vol. 216: H. Maaß, Siegel's Modular Forms and Dirichlet Series. VII, 328 pages. 1971. DM 20,-

Vol. 217: T. J. Jech, Lectures in Set Theory with Particular Emphasis on the Method of Forcing. V, 137 pages. 1971. DM 16,-

Vol. 218: C. P. Schnorr, Zufälligkeit und Wahrscheinlichkeit. IV, 212 Seiten. 1971. DM 20,-

Vol. 219: N. L. Alling and N. Greenleaf, Foundations of the Theory of Klein Surfaces. IX, 117 pages. 1971. DM 16,-

Vol. 220: W. A. Coppel, Disconjugacy. V, 148 pages. 1971. DM 16,-

Vol. 221: P. Gabriel und F. Ulmer, Lokal präsentierbare Kategorien. V, 200 Seiten. 1971. DM 18,-

Vol. 222: C. Meghea, Compactification des Espaces Harmoniques. III, 108 pages. 1971. DM 16,-

Vol. 223: U. Felgner, Models of ZF-Set Theory. VI, 173 pages. 1971. DM 16,-

Vol. 224: Revêtements Etales et Groupe Fondamental. (SGA 1). Dirigé par A. Grothendieck XXII, 447 pages. 1971. DM 30,-

Vol. 225: Théorie des Intersections et Théorème de Riemann-Roch. (SGA 6). Dirigé par P. Berthelot, A. Grothendieck et L. Illusie. XII, 700 pages. 1971. DM 40,-

Vol. 226: Seminar on Potential Theory, II. Edited by H. Bauer. IV, 170 pages. 1971. DM 18,-

Vol. 227: H. L. Montgomery, Topics in Multiplicative Number Theory. IX, 178 pages. 1971. DM 18,-

Vol. 228: Conference on Applications of Numerical Analysis. Edited by J. Ll. Morris. X, 358 pages. 1971. DM 26,-

Vol. 229: J. Väisälä, Lectures on n-Dimensional Quasiconformal Mappings. XIV, 144 pages. 1971. DM 16,-

Vol. 230: L. Waelbroeck, Topological Vector Spaces and Algebras. VII, 158 pages. 1971. DM 16,-

Vol. 231: H. Reiter, L^1-Algebras and Segal Algebras. XI, 113 pages. 1971. DM 16,-

Vol. 232: T. H. Ganelius, Tauberian Remainder Theorems. VI, 75 pages. 1971. DM 16,-

Vol. 233: C. P. Tsokos and W. J. Padgett. Random Integral Equations with Applications to stochastic Systems. VII, 174 pages. 1971. DM 18,-

Vol. 234: A. Andreotti and W. Stoll. Analytic and Algebraic Dependence of Meromorphic Functions. III, 390 pages. 1971. DM 26,-

Vol. 235: Global Differentiable Dynamics. Edited by O. Hájek, A. J. Lohwater, and R. McCann. X, 140 pages. 1971. DM 16,-

Vol. 236: M. Barr, P. A. Grillet, and D. H. van Osdol. Exact Categories and Categories of Sheaves. VII, 239 pages. 1971. DM 20,-

Vol. 237: B. Stenström, Rings and Modules of Quotients. VII, 136 pages. 1971. DM 16,-

Vol. 238: Der kanonische Modul eines Cohen-Macaulay-Rings. Herausgegeben von Jürgen Herzog und Ernst Kunz. VI, 103 Seiten. 1971. DM 16,-

Vol. 239: L. Illusie, Complexe Cotangent et Déformations I. XV, 355 pages. 1971. DM 26,-

Vol. 240: A. Kerber, Representations of Permutation Groups I. VII, 192 pages. 1971. DM 18,-

Vol. 241: S. Kaneyuki, Homogeneous Bounded Domains and Siegel Domains. V, 89 pages. 1971. DM 16,-

Vol. 242: R. R. Coifman et G. Weiss, Analyse Harmonique Non-Commutative sur Certains Espaces. V, 160 pages. 1971. DM 16,-

Vol. 243: Japan-United States Seminar on Ordinary Differential and Functional Equations. Edited by M. Urabe. VIII, 332 pages. 1971. DM 26,-

Vol. 244: Séminaire Bourbaki - vol. 1970/71. Exposés 382-399. V, 356 pages. 1971. DM 26,-

Vol. 245: D. E. Cohen, Groups of Cohomological Dimension One. V, 99 pages. 1972. DM 16,-

Vol. 246: Lectures on Rings and Modules. Tulane University Ring and Operator Theory Year, 1970-1971. Volume I. X, 661 pages. 1972. DM 40,-

Vol. 247: Lectures on Operator Algebras. Tulane University Ring and Operator Theory Year, 1970-1971. Volume II. XI, 786 pages. 1972. DM 40,-

Vol. 248: Lectures on the Applications of Sheaves to Ring Theory. Tulane University Ring and Operator Theory Year, 1970-1971. Volume III. VIII, 315 pages. 1971. DM 26,-

Vol. 249: Symposium on Algebraic Topology. Edited by P. J. Hilton. VII, 111 pages. 1971. DM 16,-

Vol. 250: B. Jónsson, Topics in Universal Algebra. VI, 220 pages. 1972. DM 20,-

Vol. 251: The Theory of Arithmetic Functions. Edited by A. A. Gioia and D. L. Goldsmith VI, 287 pages. 1972. DM 24,-

Vol. 252: D. A. Stone, Stratified Polyhedra. IX, 193 pages. 1972. DM 18,-

Vol. 253: V. Komkov, Optimal Control Theory for the Damping of Vibrations of Simple Elastic Systems. V, 240 pages. 1972. DM 20,-

Vol. 254: C. U. Jensen, Les Foncteurs Dérivés de $\varprojlim$ et leurs Applications en Théorie des Modules. V, 103 pages. 1972. DM 16,-

Vol. 255: Conference in Mathematical Logic - London '70. Edited by W. Hodges. VIII, 351 pages. 1972. DM 26,-

Vol. 256: C. A. Berenstein and M. A. Dostal, Analytically Uniform Spaces and their Applications to Convolution Equations. VII, 130 pages. 1972. DM 16,-

Vol. 257: R. B. Holmes, A Course on Optimization and Best Approximation. VIII, 233 pages. 1972. DM 20,-

Vol. 258: Séminaire de Probabilités VI. Edited by P. A. Meyer. VI, 253 pages. 1972. DM 22,-

Vol. 259: N. Moulis, Structures de Fredholm sur les Variétés Hilbertiennes. V, 123 pages. 1972. DM 16,-

Vol. 260: R. Godement and H. Jacquet, Zeta Functions of Simple Algebras. IX, 188 pages. 1972. DM 18,-

Vol. 261: A. Guichardet, Symmetric Hilbert Spaces and Related Topics. V, 197 pages. 1972. DM 18,-

Vol. 262: H. G. Zimmer, Computational Problems, Methods, and Results in Algebraic Number Theory. V, 103 pages. 1972. DM 16,-

Vol. 263: T. Parthasarathy, Selection Theorems and their Applications. VII, 101 pages. 1972. DM 16,-

Vol. 264: W. Messing, The Crystals Associated to Barsotti-Tate Groups: With Applications to Abelian Schemes. III, 190 pages. 1972. DM 18,-

Vol. 265: N. Saavedra Rivano, Catégories Tannakiennes. II, 418 pages. 1972. DM 26,-

Vol. 266: Conference on Harmonic Analysis. Edited by D. Gulick and R. L. Lipsman. VI, 323 pages. 1972. DM 24,-

Vol. 267: Numerische Lösung nichtlinearer partieller Differential- und Integro-Differentialgleichungen. Herausgegeben von R. Ansorge und W. Törnig, VI, 339 Seiten. 1972. DM 26,-

Vol. 268: C. G. Simader, On Dirichlet's Boundary Value Problem. IV, 238 pages. 1972. DM 20,-

Vol. 269: Théorie des Topos et Cohomologie Etale des Schémas. (SGA 4). Dirigé par M. Artin, A. Grothendieck et J. L. Verdier. XIX, 525 pages. 1972. DM 50,-

Vol. 270: Théorie des Topos et Cohomologie Etale des Schémas. Tome 2. (SGA 4). Dirigé par M. Artin, A. Grothendieck et J. L. Verdier. V, 418 pages. 1972. DM 50,-

Vol. 271: J. P. May, The Geometry of Iterated Loop Spaces. IX, 175 pages. 1972. DM 18,-

Vol. 272: K. R. Parthasarathy and K. Schmidt, Positive Definite Kernels, Continuous Tensor Products, and Central Limit Theorems of Probability Theory. VI, 107 pages. 1972. DM 16,-

Vol. 273: U. Seip, Kompakt erzeugte Vektorräume und Analysis. IX, 119 Seiten. 1972. DM 16,-

Vol. 274: Toposes, Algebraic Geometry and Logic. Edited by. F. W. Lawvere. VI, 189 pages. 1972. DM 18,-

Vol. 275: Séminaire Pierre Lelong (Analyse) Année 1970-1971. VI, 181 pages. 1972. DM 18,-

Vol. 276: A. Borel, Représentations de Groupes Localement Compacts. V, 98 pages. 1972. DM 16,-

Vol. 277: Séminaire Banach. Edité par C. Houzel. VII, 229 pages. 1972. DM 20,-